Conservation and Ecology of Raptors

Conservation and Ecology of Raptors

Editors

Vladimir D. Dobrev
Dobromir D. Dobrev
Dimitar A. Demerdzhiev

MDPI

Basel • Beijing • Wuhan • Barcelona • Belgrade • Novi Sad • Cluj • Manchester

Editors
Vladimir D. Dobrev
Bulgarian Society for the Protection of Birds
Plovdiv, Bulgaria

Dobromir D. Dobrev
Bulgarian Society for the Protection of Birds
Plovdiv, Bulgaria

Dimitar A. Demerdzhiev
Bulgarian Society for the Protection of Birds
Plovdiv, Bulgaria

Editorial Office
MDPI
St. Alban-Anlage 66
4052 Basel, Switzerland

This is a reprint of articles from the Special Issue published online in the open access journal *Diversity* (ISSN 1424-2818) (available at: https://www.mdpi.com/journal/diversity/special_issues/Raptors).

For citation purposes, cite each article independently as indicated on the article page online and as indicated below:

Lastname, A.A.; Lastname, B.B. Article Title. *Journal Name* **Year**, *Volume Number*, Page Range.

ISBN 978-3-0365-8692-2 (Hbk)
ISBN 978-3-0365-8693-9 (PDF)
doi.org/10.3390/books978-3-0365-8693-9

Cover image courtesy of Dobromir Dobrev

Contents

Preface

Raptors are a diverse avian guild distributed worldwide and adapted to different environments. At present, this guild of birds faces a high risk of extinction due to numerous threats that occur in their breeding, migration and wintering areas. Understanding and studying their demography, abundance, diet, resource selection, movements, ecological requirements, adaptive capacity and major threats will provide valuable information about current ecosystem functioning and relevant conservation measures. This Special Issue provides new insights into these topics and demonstrates different survey techniques that researchers can use to study them.

Vladimir D. Dobrev, Dobromir D. Dobrev, and Dimitar A. Demerdzhiev
Editors

diversity

MDPI

Editorial

Conservation and Ecology of Raptors: Introduction to the Special Issue

Dobromir Dobrev [1,*], Vladimir Dobrev [1] and Dimitar Demerdzhiev [1,2]

1 Bulgarian Society for the Protection of Birds/BirdLife Bulgaria, 5, Leonardo da Vinci Str., 4000 Plovdiv, Bulgaria; vd.dobrev@gmail.com (V.D.); dimitar.demerdzhiev@bspb.org or dimitar.demerdzhiev@gmail.com (D.D.)

2 National Museum of Natural History, Bulgarian Academy of Sciences, 1000 Sofia, Bulgaria

* Correspondence: dobromir.dobrev@bspb.org or dobromir.dobrev1@gmail.com

Raptors are a diverse avian guild distributed worldwide and adapted to different environments. They have attracted a strong interest from scientists because of their conservation status and ecology. Raptors provide critical ecosystem services based on their role in natural food chains. They have evolved ecologically to specialize and adapt to different habitats and food resources. Thus, they are subject to a wide range of biological, ecological, and evolutionary studies. At present, this guild of birds faces a high risk of extinction due to numerous threats that occur in their breeding, migration, and wintering areas. Globally, 18% of raptors are threatened with extinction and 52% have declining populations. They stand at the top of the trophic pyramids and are normally distributed at low densities over vast areas. Some species are long-distance migrants while others are resident. Birds of prey play a crucial role in the conservation of ecosystems and are assumed as flagship species in relation to other taxa. Understanding and studying their demography, abundance, diet, resource selection, movements, ecological requirements, adaptive capacity, and the threats they face, will provide valuable information about the current ecosystem functioning and status.

The importance of raptors is even more pronounced, since by protecting them and their habitats, we are actually preserving a variety of other invertebrate and vertebrate species. At a global scale, many raptors are considered umbrella species of particular importance in conservation. By studying their intra- and inter-specific interactions, we can learn more about their diet, competition, and movement ecology, as well as the diversity of environments they inhabit. These environments, on other hand, harbor diverse communities of prey species, which are utilized by raptors. However, raptors' behavior may differ between different age classes and is very often determined by sex depending on the species. Since raptor population trends are mostly driven by the given species' demography, turnover, breeding rates, and human-induced mortality, studying these aspects of their ecology is fundamental to the conservation interventions to be conducted.

The twelve papers in this Special Issue address all these topics.

Four papers contribute to understanding prey composition and adaptations to different food sources as well as the related spatial and temporal limitations [1–4]. These studies were conducted in different regions of Europe—two in Bulgaria [1,2], one in Poland [5], and one in Spain [3]. The two studies from Bulgaria were based on a large apex predator—the eastern imperial eagle, and they have revealed the unknown and understudied features of the species' foraging behavior and diet variations. In the study from Poland, the authors aimed to compare the diet of the white-tailed eagle in optimal and suboptimal conditions to investigate how population development affects prey composition there. The study suggested that eagles cope with a lack of optimal prey by ranging farther and exploring non-optimal foraging habitats. The last study of this group investigated the patterns of use of household waste by overwintering red kites in southeastern Madrid, central Spain. The

Citation: Dobrev, D.; Dobrev, V.; Demerdzhiev, D. Conservation and Ecology of Raptors: Introduction to the Special Issue. *Diversity* **2023**, *15*, 889. https://doi.org/10.3390/d15080889

Received: 20 July 2023
Accepted: 22 July 2023
Published: 27 July 2023

results reveal the constant presence of relatively low numbers of kites during the whole winter period in the studied landfill. The Special Issue continues with an overview of nest site selection in three apex predators, which are positioned at the highest trophic level of the food web, being the best indicators of the ecosystem's richness and functioning [4,6]. In the first study [4], species distribution modelling was applied to study the association of the golden eagle and Bonelli's eagle to different natural and anthropogenic factors to distinguish both species' niche separation and coexistence. The other study conducted a multi-scale model analysis of eagle owl territories in Israel to understand the habitat requirements of the species. Interestingly, the authors found that the species' distribution was limited by the availability of quality nest locations, i.e., quarries and caves.

Some of the most important traits in raptor breeding cycles and territory occupancy are the turnover rates, dispersal, and prey species richness. Four of the Special Issue papers give an overview of the pertinent aspects of this topic [7–10] and reveal important findings concerning territory occupancy. A study from Finland [7] used genetic sampling to understand the genetic diversity and population structure of the golden eagle. The authors examined the turnover rate using chick genotypes and revealed that the turnover rate of this population was 23%. Another study from Scotland [8] sought to find and analyze pre-settlement data from 37 GPS-tagged nestling golden eagles, later recorded as having settled in their first territory. The third of this set of studies analyzed home range size and space use in territorial Bonelli's eagles in Spain [9]. Authors found that the home range size of females was slightly smaller than that of males, due to the decreased activity during the breeding season as a result of egg laying, incubation, and nest attendance. The last study of this series [10] investigated the activity patterns of goshawks and their role as an indicator of bird abundance. One of the studies [11] in this Special Issue aimed to determine how changes (caused by external agents or by endogenous modifications) to the species interacting with the lesser kestrel could induce changes (both positive and negative) to its population stock. The authors used a qualitative modelling approach to identify the species (predators, competitors, prey) that interacted with the species of interest (the lesser kestrel) in southern Italy, mapped all of the ecological interactions among these species, and simulated different management strategies that could increase kestrel population stock by targeting the species that interacted with it. Finally, the last of the studies dealt with pure ecological research, investigating whether road kills were density-dependent [12]. The study proved that barn owl road kills were density-dependent and demonstrated the importance of monitoring breeding and population numbers in roadkill studies.

The papers in this Special Issue address many of the most essential questions related to the biology, ecology, and conservation of raptors. Cumulatively, the papers in this Special Issue provide important answers to many of these questions and establish a foundation for further research on these topics.

Conflicts of Interest: The authors declare no conflict of interest.

References

1. Demerdzhiev, D.; Angelov, I.; Dobrev, D. Foraging Patterns of Non-Territorial Eastern Imperial Eagle (*Aquila heliaca*): A Case of Successful Adaptation. *Diversity* **2022**, *14*, 1060. [CrossRef]
2. Demerdzhiev, D.; Boev, Z.; Dobrev, D.; Nedyalkov, N.; Petrov, T. Does Temporal and Spatial Diet Alteration Lead to Successful Adaptation of the Eastern Imperial Eagle, a Top Predator? *Diversity* **2022**, *14*, 1000. [CrossRef]
3. Vicente-Hernández, Í.; Martinez, F.; Blanco, G. Rabbits or Refuse? Landfill Use and Relevance as a Food Source for an Increasing Wintering Population of the Red Kite. *Diversity* **2023**, *15*, 704. [CrossRef]
4. Solanou, M.; Trypidaki, E.; Georgopoulou, E.; Damianakis, K.; Kardamaki, A.; Xirouchakis, S. Selection of Nesting Habitat and Insular Niche Separation of Two Sympatric Aquila Species. *Diversity* **2022**, *14*, 1136. [CrossRef]
5. Mirski, P.; Komar, E. The White-Tailed Eagle, the Apex Predator, Adjusts Diet towards Larger Prey in Suboptimal Territories. *Diversity* **2023**, *15*, 747. [CrossRef]
6. Hadad, E.; Malkinson, D.; Yosef, R.; Weil, G.; Charter, M. Importance of Mesohabitat for Nest-Site Selection in Breeding Eagle Owls (*Bubo bubo*): A Multi-Scale Model. *Diversity* **2022**, *14*, 438. [CrossRef]
7. Kylmänen, A.; Karabanina, E.; Ollila, T.; Ponnikas, S.; Kvist, L. Turnover and Natal Dispersal in the Finnish Golden Eagle (*Aquila chrysaetos*) Population. *Diversity* **2023**, *15*, 567. [CrossRef]

8. Fielding, A.; Anderson, D.; Benn, S.; Reid, R.; Tingay, R.; Weston, E.; Whitfield, D. Substantial Variation in Prospecting Behaviour of Young Golden Eagles Aquila chrysaetos Defies Expectations from Potential Predictors. *Diversity* **2023**, *15*, 506. [CrossRef]
9. Morollón, S.; Urios, V.; López-López, P. Home-range Size and Space Use of Territorial Bonelli's Eagles (*Aquila Fasciata*) Tracked by High Resolution GPS/GSM Telemetry. *Diversity* **2022**, *14*, 1082. [CrossRef]
10. Väli, Ü.; Grosberg, J.; Mellov, P.; Tali, T.; Mirski, P. Is the Northern Goshawk an Efficient Bioindicator of Avian Abundance and Species Richness in Urban Environments? *Diversity* **2023**, *15*, 749. [CrossRef]
11. Ferrarini, A.; Gustin, M.; Pellegrino, S.; Giglio, G. A Community-Level Approach to Species Conservation: A Case Study of Falco naumanni in Southern Italy. *Diversity* **2022**, *14*, 566. [CrossRef]
12. Meyrom, K.; Yosef, R.; Charter, M. Are Roadkills Density-Dependent? Case Study of the Barn Owl (*Tyto alba*). *Diversity* **2023**, *15*, 412. [CrossRef]

diversity

MDPI

Article

Is the Northern Goshawk an Efficient Bioindicator of Avian Abundance and Species Richness in Urban Environments?

Ülo Väli [1,*], Jaan Grosberg [1], Pelle Mellov [1], Tiiu Tali [1] and Paweł Mirski [1,2]

1 Institute of Agricultural and Environmental Sciences, Estonian University of Life Sciences, Kreutzwaldi 5D, 51006 Tartu, Estonia
2 Faculty of Biology, University of Białystok, Ciołkowskiego 1J, 15-245 Białystok, Poland
* Correspondence: ulo.vali@emu.ee

Abstract: Monitoring of biodiversity in expanding urban areas is an essential part of wildlife conservation. There is evidence that raptors, such as Northern Goshawks (*Accipiter gentilis*), are effective bioindicator species in urban areas, however, their relationship with other bird populations is not clearly established. We asked whether activity patterns of Goshawks are a reliable indicator of wintering bird abundance and diversity in urban ecosystems. We tracked the movement of eight GPS-tagged Goshawks in the city of Tartu (Estonia) and analysed the numbers and diversity of birds in the same area using direct mapping and occasional data obtained from birdwatchers. The direct mapping approach revealed that the number of birds and avian species richness were higher in Goshawk activity hotspots than at random sites in 2022, however, no such differences were detected in 2023. Analysis of occasional citizen-collected data showed no effect of avian abundance nor species richness on the distribution of Goshawk activity. These results suggested that the movements of Goshawks may indicate the abundance and diversity of its prey, however, this relationship depends on the detection methodology. Hence, raptors are a promising bioindicator in urban environments, but results should be interpreted with caution, particularly when using citizen-collected data.

Keywords: bioindicator species; bird abundance; bird diversity; birds of prey; citizen science; GPS telemetry; raptors; sentinel species; urban biodiversity

Citation: Väli, Ü.; Grosberg, J.; Mellov, P.; Tali, T.; Mirski, P. Is the Northern Goshawk an Efficient Bioindicator of Avian Abundance and Species Richness in Urban Environments? *Diversity* **2023**, *15*, 749. https://doi.org/10.3390/d15060749

Academic Editors: Vladimir D. Dobrev, Dobromir D. Dobrev, Dimitar A. Demerdzhiev, Miguel Ferrer and Alan Feduccia

Received: 30 April 2023
Revised: 31 May 2023
Accepted: 5 June 2023
Published: 7 June 2023

1. Introduction

Urbanised areas have become the most rapidly expanding habitat type worldwide [1] and urbanisation is one of the main threats to biodiversity [2–4]. However, a number of wildlife species have adapted to urban environments. Hence, preserving and monitoring biodiversity in human-dominated areas are becoming essential parts of maintaining biodiversity on the global scale [5].

Assessing total biodiversity is laborious and costly. Therefore, it is often evaluated using bioindicators, which are species or assemblages of species reactive to environmental changes [6]. Birds, for example, are highly visible and sensitive to changes in habitat structure and composition, therefore, they are excellent indicators of habitat quality, including that in urban environments [3,5,7]. However, comprehensive avifaunal inventories are often not feasible. Thus, well-chosen bioindicator species or species groups may be an efficient shortcut to evaluate ecosystem quality [8]. For example, large predators, raptors in particular, are considered good indicators of viable ecosystems [9,10]. Indeed, there is accumulating evidence that various raptor species are efficient surrogates for biodiversity in various ecosystems [11–14].

The Northern Goshawk *Accipiter gentilis* (hereafter Goshawk) is a flexible avian apex predator inhabiting various landscapes. Primarily, Goshawk is a forest-dwelling species, however, it also thrives in mosaic agricultural landscapes and has recently colonised cities [15]. Therefore, this species has been used as an indicator of biodiversity in forests [12,13], farmland [16] and urban areas [17,18]. Goshawks forage primarily on birds [15]. As the efficacy of a bioindicator is higher for taxa with a stronger ecological connection to the predator [10], Goshawk distributions are expected to effectively indicate avian abundance and diversity.

Northern bird populations, including Goshawk populations, are strongly limited by the occurrence of prey during winter [19–21]. Under harsh conditions, many birds inhabit areas in proximity to humans [19,22,23] and may even move to cities from less populated areas [24]. Goshawks, in turn, may follow the movements of their prey between habitats [15]. Hence, cities attract wintering hawks and, in addition to local residents, nonbreeding individuals may be concentrated in these areas. This situation provides an excellent chance to directly study relationships between predators and prey because associations with nests, which bias spatial behaviour, are limited or lacking. Earlier, Natsukawa [18] found that Goshawk nest site selection in a city corresponded to the habitat selection of wintering birds, indicating that Goshawk nest sites may serve as a surrogate for hotspots of avian diversity in urban environments. However, as these previous data sets were temporally separated, the direct link between Goshawks and other birds remains untested.

The past few decades have witnessed the emergence and growth of several new scientific methods. First, several novel technologies, such as GPS-based telemetry, have seen rapid advances. Movement ecology, owing to rapid advances in telemetry technologies, is an active field of research with great potential for investigations of broad, biodiversity-scale issues [25]. This enables the replacement of landscape-level correlations with the actual pinpointing of activity centres of animals. Second, citizen science (i.e., the involvement of non-scientists in data collection for scientific research) has been expanding, in part owing to technological developments [26–28]. Citizen science provides an opportunity to conduct research at broad spatial scales, which are impossible to sample extensively using traditional field research models [29,30]. Citizen scientists, for example, collect field data related to species distributions and abundance [27,29,31]. Extensive datasets based on opportunistic observations by amateurs have contributed to faunistic surveys and correlative ecological analyses [29,31,32].

The aim of this study was to test whether Goshawk habitat use is related to the distribution of wintering birds in an urban environment. In particular, we tracked movements of eight GPS-tagged Goshawks in the city of Tartu and analysed the number and diversity of birds in the same area. We hypothesised that the activity centres of Goshawks are positively associated with avian abundance and species richness. We explored the abundance and diversity of birds in two ways. First, we mapped birds in sites preferred by GPS-tracked Goshawks and in control sites; second, we analysed occasional observations of birdwatchers. Hence, by comparison of the results obtained using the two approaches, our findings provide insight into the utility of citizen science for estimating avian abundance and diversity in urban environments.

2. Materials and Methods

This study was conducted in Tartu, Estonia, in north-eastern Europe (58°23′ N 26°43′ E). Tartu is the second largest city in Estonia with a population of c. 100,000 people. The average annual air temperature is 6.1 °C and the coldest month is February (on average −5.3 °C [33]). Tartu has rather diverse land use [34], with the dominant features being residential areas (covering 30.7% of the area), open green areas (28.7%) and roads (20.4%). Afforested areas (9.0%), open lands without vegetation (4.3%) and cultivated lands (3.5%) cover smaller portions of the landscape. Wetlands and water bodies, such as the river Emajõgi passing through the city, hold significant ecological value despite occupying a minor proportion of the area, accounting for 2.7% and 0.7% of the landscape, respectively (Figure 1).

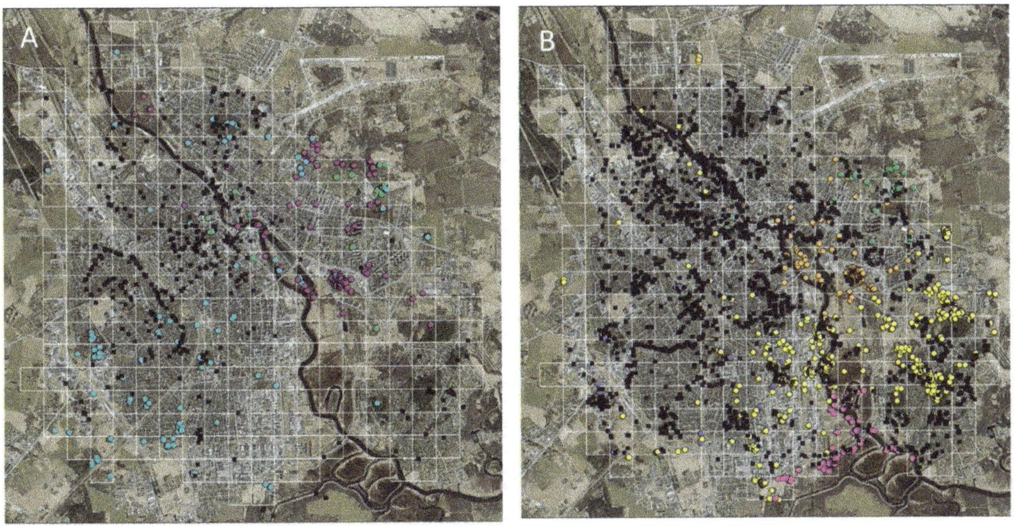

Figure 1. Citizen-collected occasional bird observations (black squares) and registered locations of Goshawks (circles, where individuals are shown in different colours) in (**A**) February 2022 and (**B**) November–February 2022/2023.

The study was conducted in two winters, in 2021/2022 (February 2022) and 2022/2023 (November 2022–February 2023). In total, seven GPS-tagged Goshawks (six males and one female) were included in the study (Table 1). Each bird was equipped with a 15–30 g (<3% of the body mass) solar-powered GSM/GPRS logger (UAB Ornitela, Vilnius, Lithuania) as a backpack using Teflon harnesses. Seven birds were followed during one winter but an adult male provided data in both study winters. All birds were followed for the entire study periods, i.e., for 28 days in 2021/2022 (84 tracking days in total) and for 120 days in 2022/2023 (600 tracking days in total). However, the datasets varied owing mainly to limited light in the winter, preventing loggers from recharging. Eventually, we used 491 Goshawk locations from 2021/2022 and 1304 locations from 2022/2023 (Table 1).

Table 1. Age, sex, tracking period and number of GPS-fixes of tracked Goshawks.

Logger No.	Age	Sex	Tracking Winter	No. of GPS-Fixes
171095	Adult	Male	2021/2022	47
190723	Immature	Male	2021/2022	205
190725	Immature	Male	2021/2022	239
171095	Adult	Male	2022/2023	60
190703	Adult	Male	2022/2023	146
190728	Adult	Male	2022/2023	30
212340	Adult	Male	2022/2023	853
212347	Adult	Female	2022/2023	215

The abundance and distribution of wintering birds in the city of Tartu was determined using two approaches. First, the authors (J.G., P.Me., T.T., and Ü.V.) mapped the birds on 14 to 20 February 2022 and on 14 to 20 February 2023 (Table 2). The city of Tartu was divided into 400 × 400 m squares (Figure 1). Out of 299 squares, 50 squares at town edges that contained >60% of land outside the borders of Tartu and nine squares that were highly (>60%) afforested and were not classified as urban were excluded. The remaining 240 grid squares were overlaid with GPS-telemetry data for Goshawks to select two independent sets (one for each season) of Goshawk activity hotspots and random squares. The hotspots were defined as the 25 grid squares with the highest number of Goshawk GPS-fixes in the given season. To avoid clustering, we selected only the squares with highest number of Goshawk locations and omitted all bordering squares (sharing a corner was allowed). To compare sites used by Goshawks with available urban sites, another 25 squares were randomly drawn from those that were not used by Goshawks. Eventually, only five hotspot squares and three random points were repeatedly selected in the two seasons; additionally one random point from 2022 was a hotspot in 2023. In 2022/2023, most hotspots were consistent throughout the winter (Figure 2).

Table 2. Total numbers of bird individuals and species counted via direct mapping and recorded occasionally by birdwatchers.

Year	Square Type	All Birds		Medium-Sized Birds		"Local" Birds	
		Abundance	Species Richness	Abundance	Species Richness	Abundance	Species Richness
			Mapping data				
2022	Hotspot	1636	30	746	12	1505	29
2022	Random	1508	27	935	11	1399	27
2023	Hotspot	1513	35	631	14	1413	33
2023	Random	1815	28	580	11	1704	26
			Occasional data				
2022		7946	56	5365	23		
2023		17299	69	10382	28		

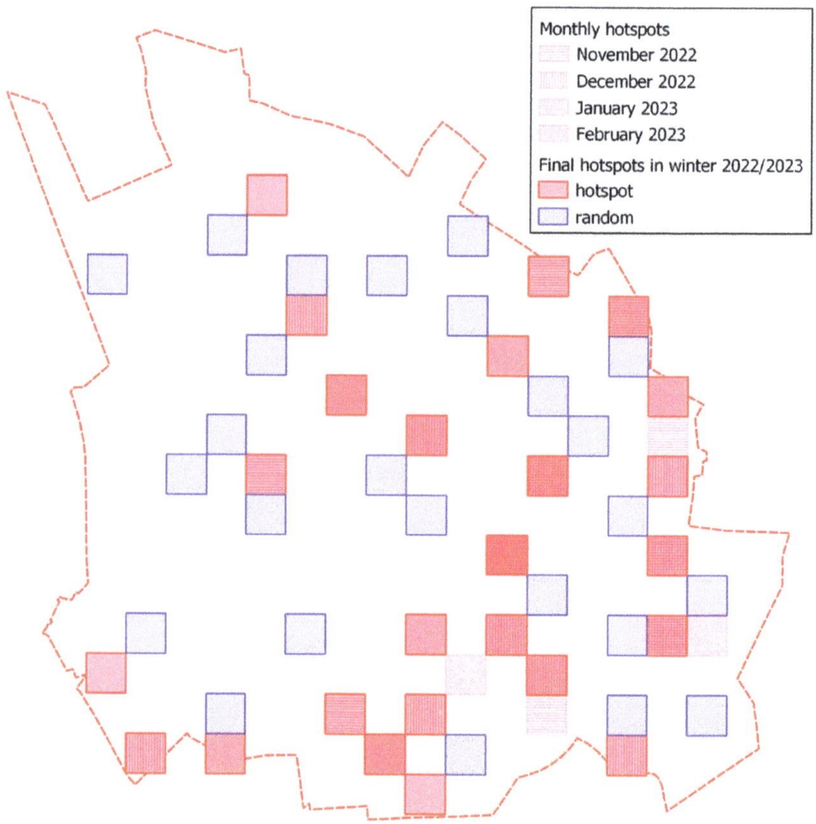

Figure 2. Distribution and consistency of Goshawk activity hotspots and the distribution of control plots in the winter 2022/2023.

Second, we used the data from citizen scientists deposited in PlutoF [35], a portal incorporating observations of Estonian birdwatchers. In early February 2022 and early November 2023, calls were published on social media platforms to encourage bird enthusiasts to collect observations in Tartu and deposit these in the PlutoF database. Collected occasional bird data (in February 2022 and November 2022–February 2023) were analysed using the same grid used in the first approach. We attempted to avoid two potential methodological caveats. First, the study area was not uniformly covered by the bird observations, nor by the home ranges of Goshawks. To avoid the effect of spatial non-overlap of the two data sets, we included only squares with at least one bird sighting and at least one Goshawk record in the analyses. Secondly, the same observers may have visited the same squares repeatedly. To avoid the cumulative effect of repeated visits, only the maximum number of each bird species in each square was included.

In each square, species richness and the abundance of each bird species were calculated. All bird species were included in initial analyses. Thereafter, only medium-sized birds (ducks, pigeons, most corvids, thrushes, etc.) were included as potential prey items for Goshawks. In the analysis of mapping data, the effect of "local" birds identified as potential prey in a given location was analysed separately (i.e., birds flying over were excluded). Owing to the limits of data deposition in the PlutoF database, the latter specification was not possible in the analysis of occasional data.

The bird mapping data were analysed using logistic regression models, where grid square type was a binary response variable, and avian abundance or species richness were covariates. Owing to the strong collinearity, abundance and species richness were analysed via separate models. In the analysis of occasional observations, we used linear models where the number of Goshawk GPS-fixes was a continuous response variable; again, avian abundance or species richness were covariates. Initial models included factor year and its interaction with covariates but final models were developed for each year separately. All continuous variables were log-transformed prior to analyses.

3. Results

The total number of species, but not the abundance, was always higher in squares with high Goshawk activity (hotspots) than in random squares (Table 2). According to the logistic regression analysis of bird mapping data, bird abundance was nearly significantly higher in Goshawk activity hotspots than in random squares, and the effect of year was also nearly significant (Table 3). In 2022, there were more birds in Goshawk hotspots (i.e., grid squares with high Goshawk activity) than in random squares, however, no such difference was detected in 2023 (Figure 3). Avian species richness had a nearly significant effect on the distribution of Goshawk activity and its interaction with year had a similar effect (Table 3); species richness was significantly higher at Goshawk hotspots in 2022 but not in 2023 (Figure 3). Similar tendencies were detected for the abundances (2022: $t = 1.79$, $p = 0.081$; 2023: $t = 0.54$; $p = 0.59$) and species richness (2022: $t = 1.58$, $p = 0.121$; 2023: $t = 0.43$; $p = 0.672$) of 'local' birds (Table 3). However, the abundances (2022: $t = 0.84$, $p = 0.631$; 2023: $t = 0.78$; $p = 0.438$) or species richness of medium-sized birds had no effect on Goshawk activity (2022: $t = 1.07$, $p = 0.292$; 2023: $t = 0.14$; $p = 0.886$; Table 3).

Table 3. Logistic regression models describing the effect of avian abundance and species richness (both variables log-transformed) on grid square type (Goshawk activity hotpots vs. random squares).

Variable	Estimate	SE	t	p
	All birds			
Intercept	−1801.1	1051.1	−1.71	0.090
Abundance	440.1	262.5	1.68	0.097
Year	0.9	0.5	1.71	0.090
Abundance × Year	−0.2	0.1	−1.68	0.097
Intercept	−1680.5	945.0	−1.78	0.079

Table 3. *Cont.*

Variable	Estimate	SE	t	*p*
Species richness	593.4	335.7	1.77	0.080
Year	0.8	0.5	1.78	0.079
Species richness × Year	−0.3	0.2	−1.77	0.081
	'Local' birds			
Intercept	−1553.7	978.9	−1.59	0.116
Abundance	388.3	250.0	1.55	0.124
Year	0.8	0.5	1.59	0.116
Abundance × Year	−0.2	0.1	−1.55	0.124
Intercept	−1051.2	834.4	−1.26	0.211
Species richness	381.3	306.2	1.25	0.216
Year	0.5	0.4	1.26	0.211
Species richness × Year	−0.2	0.2	−1.25	0.216
	Medium-sized birds			
Intercept	−4.2	532.6	−0.01	0.994
Abundance	−37.9	180.0	−0.21	0.834
Year	0.0	0.3	0.01	0.993
Abundance × Year	0.0	0.1	0.21	0.834
(Intercept)	−495.1	515.8	−0.96	0.340
Species richness	245.6	284.7	0.86	0.391
Year	0.2	0.3	0.96	0.339
Species richness × Year	−0.1	0.1	−0.86	0.391

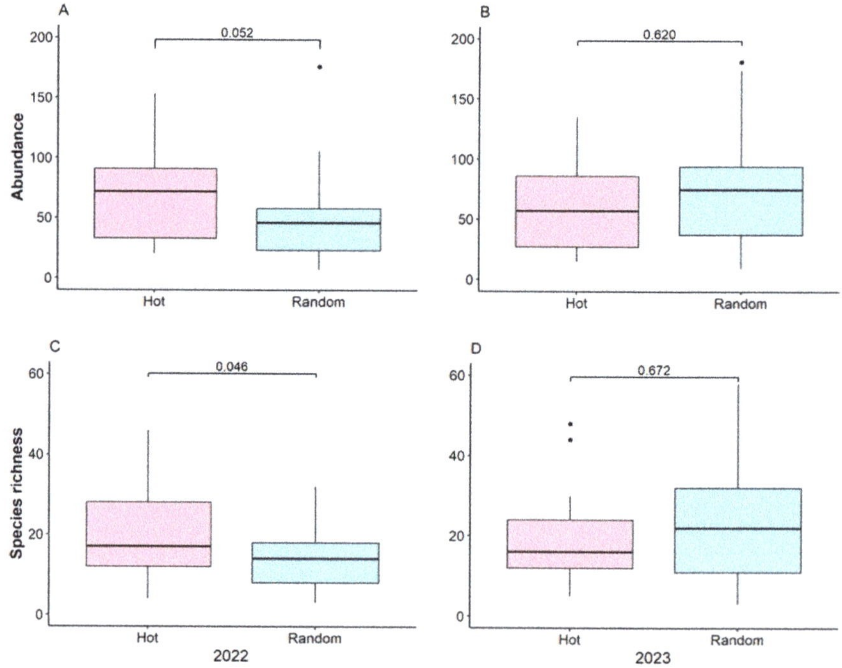

Figure 3. (**A**,**B**) Mapping-based abundance and (**C**,**D**) species richness of wintering birds in 2022 (**A**,**C**) and in 2023 (**B**,**D**) in the grid squares with high Goshawk activity (hotspots) and in random squares in Tartu. The bold line indicates the median, the box shows quartiles, the whiskers indicate the extreme data points within 1.5× the interquartile range from the quartile boundaries and dots are data points beyond that range. *p*-values for univariate logistic regression models are indicated in brackets.

In the analysis of occasional data, we did not detect an effect of total avian abundance ($F_{3,130} = 2.1$, $p = 0.101$) or species richness ($F_{3,130} = 2.0$, $p = 0.121$) on the number of Goshawk fixes in grid squares (Table 4). Additionally, there was no significant interaction with year (Table 4). Similarly, we did not detect any effects when years were analysed separately (Figure 4). We did not detect an effect of bird abundance ($F_{3,114} = 2.1$, $p = 0.102$; 2022: $F_{1,28} = 0.04$, $p = 0.836$; 2023: $F_{1,86} = 1.02$, $p = 0.315$) or richness ($F_{3,114} = 2.2$, $p = 0.097$; 2022: $F_{1,28} = 0.09$, $p = 0.771$; 2023: $F_{1,116} = 0.59$, $p = 0.445$) when only medium-sized birds were included in the analysis.

Table 4. Linear regression models describing the effect of avian abundance and species richness (both variables log-transformed) on Goshawk activity (number of GPS-fixes in grid squares).

Variable	Estimate	SE	t	*p*
	All birds			
Intercept	0.3	0.3	1.26	0.209
Abundance	0.0	0.1	0.00	0.997
Year	0.4	0.3	1.31	0.194
Abundance × Year	−0.1	0.2	−0.46	0.643
Intercept	−516.2	412.8	−1.25	0.213
Species richness	−39.8	558.3	−0.07	0.943
Year	0.3	0.2	1.25	0.213
Species richness × Year	0.0	0.3	0.07	0.943
	Medium-sized birds			
Intercept	0.3	0.2	1.33	0.185
Abundance	0.0	0.1	0.15	0.879
Year	0.4	0.3	1.61	0.109
Abundance × Year	−0.1	0.2	−0.71	0.480
Intercept	−799.3	452.2	−1.77	0.080
Species richness	624.1	978.5	0.64	0.525
Year	0.4	0.2	1.77	0.080
Species richness × Year	−0.3	0.5	−0.64	0.525

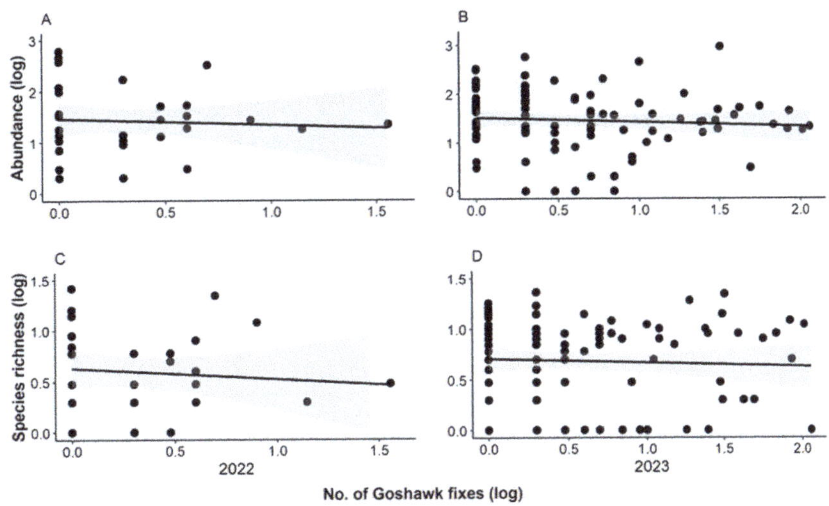

Figure 4. (**A**,**B**) Citizen-collected abundance and (**C**,**D**) species richness of wintering birds in 2022 (**A**,**C**) and in 2023 (**B**,**D**) in grid squares of Tartu in relation to the number of registered Goshawk locations.

4. Discussion

We used two different approaches to study associations between Goshawk and its prey in urban environments. In the first approach, via direct mapping, we detected a positive association in one winter but not in another. In the second approach, using occasional observations of birdwatchers, we did not detect associations between these parameters.

To address the limitations of short-term studies, we conducted this study over two winters. Variations across years may reflect the effects of weather or other features of particular winters. Furthermore, the results might have been affected by the different period of tracking and the different number (and age) of the tracked birds. However, such effects would have been detected consistently using both approaches whereas we detected differences between years only in our own mapping-based inventories but not in the analysis of citizen-collected data. This suggests that methodological differences influenced our results. Notably, total species richness (but not abundance) in both study winters was higher in Goshawk activity hotspots than in control plots.

Data for avian abundance and distribution collected by citizen scientists did not show any association with Goshawk activity centres in the first study year, which is different from the results of our mapping analysis. The citizen-collected data were rather limited in the first study winter and a substantial amount of information had to be discarded owing to the restricted spatial distribution and lack of spatial overlap with tracking data. Citizen science has other limitations, including the limited skills of participants and biases related to data collection [28,29], which could explain the conflicting results obtained via the two approaches. Evaluations of these limitations are beyond the scope of our paper, however, we stress that citizen-collected data should be analysed with caution and, if possible, results should be validated using another methodology.

Our mapping approach indicated that bird abundance and richness were significantly higher in Goshawk activity centres than in random plots in the first study season but not in the second season. The dataset for 2021/2022 was limited to late winter (i.e., February). The study period in 2021/2022 was temporally restricted and the detected association indicated a direct spatial link between Goshawk individuals and prey. In the next winter, Goshawk data were collected for 3 months, from the beginning of November to early February, and the spatial distribution of activity centres was therefore broader. Although most of the detected hotpots were the same throughout the winter, bird mapping in February may have not fully represented associations in earlier months. It is unclear why medium-sized birds, which are preferable prey for Goshawks, had no effect on its activity. The most plausible explanation is the substantially smaller sample size of this group.

Raptors are well-known indicators of biodiversity and viable ecosystems; prioritisation of conservation efforts based on their occurrence is likely to provide broad ecosystem benefits [10]. However, the efficiency of raptors as biodiversity indicators has been criticised owing to inconsistent results [36–38]. Our study, using two different approaches, suggests that conflicting results can be explained, at least in part, by methodological differences.

Raptors have been used as bioindicators at different spatial scales. On one hand, nest sites of raptors often indicate biodiversity at the microhabitat level by indirect non-causative links. For example, Goshawk nests built in diverse old-growth forest stands rich in diverse taxa, such as trees, wood-decaying fungi and butterflies [11–13]. Breeding sites of Goshawks could also serve as a useful conservation surrogate for the species richness and functional diversity of wintering birds [18]. However, this association is only correlational and it may be weaker when habitat selection by raptors differs from that of other birds [18]. On the other hand, foraging activity connects raptors directly with taxa at lower trophic levels. As many raptors cover long distances or use spatially distant sites while foraging, their movement and presence/absence data indicate ecosystem quality at the landscape (macrohabitat) scale [16,39]. However, in addition to the distribution of prey, which is determined by habitat suitability, other environmental factors, such as weather or wind conditions and the distribution of perching sites, shape the distribution of raptors [40–42]. Furthermore, intra-specific interactions, such as competition and territorialism, should be considered in data analyses. In our study, untracked Goshawks may have held territories in the western part of the town, preventing foraging by tracked Goshawks in this area.

5. Conclusions

Our data suggested that Goshawk movement patterns are potential indicators of the abundance and diversity of prey, however, the results depended on the methodological approach and should be validated in a longer survey. We emphasise that relatively costly GPS tracking can hardly be suggested as a method for bioindication; instead, information on Goshawk (or other predators') activity centres may be collected via observations by citizen scientists. Although citizen science is a promising source of data for scientific research and conservation purposes, inconsistency in data acquisition may limit its use. Our results support the view that the employment of predators as bioindicators is justified but the interpretation of results requires appropriate caution [10].

Author Contributions: Conceptualisation, Ü.V.; methodology, Ü.V.; validation, Ü.V. and P.M. (Paweł Mirski); formal analysis, Ü.V.; investigation, Ü.V., J.G., P.M. (Pelle Mellov) and T.T.; resources, Ü.V.; data curation, Ü.V.; writing—original draft preparation, Ü.V.; writing—review and editing, P.M. (Paweł Mirski); visualisation, Ü.V. and P.M. (Paweł Mirski); project administration, Ü.V.; supervision, Ü.V.; funding acquisition, Ü.V. All authors have read and agreed to the published version of the manuscript.

Funding: This research was funded by the Estonian Environmental Investments Fund, grant number 18473 and by the Estonian University of Life Sciences, grant number P180271.

Institutional Review Board Statement: Not applicable.

Data Availability Statement: The data collected in bird inventories and occasional records are deposited in the PlutoF data repository and available at https://plutof.ut.ee/ (accessed on 1 April 2023). The movement data of raptors is deposited in the Movebank data repository https://www.movebank.org/ (accessed on 1 April 2023).

Acknowledgments: We are grateful to Urmas Abel, Madis Leivits and Urmas Sellis who helped to trap and tag the most informative Goshawk in 2019, and to Anni Miller and Lisell Toomla who assisted us in bird mapping. Our sincere thanks go also to all birdwatchers whose observations made the current study possible.

Conflicts of Interest: The authors declare no conflict of interest.

References

1. Grimm, N.B.; Faeth, S.H.; Golubiewski, N.E.; Redman, C.L.; Wu, J.; Bai, X.; Briggs, J.M. Global Change and the Ecology of Cities. *Science* **2008**, *319*, 756–760. [CrossRef]
2. Miller, J.R.; Hobbs, R.J. Conservation Where People Live and Work. *Conserv. Biol.* **2002**, *16*, 330–337. [CrossRef]
3. Shochat, E.; Lerman, S.B.; Anderies, J.M.; Warren, P.S.; Faeth, S.H.; Nilon, C.H. Invasion, Competition, and Biodiversity Loss in Urban Ecosystems. *BioScience* **2010**, *60*, 199–208. [CrossRef]
4. Marzluff, J.M. Worldwide Urbanization and Its Effects on Birds. In *Avian Ecology and Conservation in an Urbanizing World*; Marzluff, J.M., Bowman, R., Donnelly, R., Eds.; Springer: Boston, MA, USA, 2001; pp. 19–47. ISBN 978-1-4615-1531-9.
5. Savard, J.-P.L.; Clergeau, P.; Mennechez, G. Biodiversity Concepts and Urban Ecosystems. *Landsc. Urban Plan.* **2000**, *48*, 131–142. [CrossRef]
6. Paoletti, M.G. Using Bioindicators Based on Biodiversity to Assess Landscape Sustainability. *Agric. Ecosyst. Environ.* **1999**, *74*, 1–18. [CrossRef]
7. Faeth, S.H.; Bang, C.; Saari, S. Urban Biodiversity: Patterns and Mechanisms. *Ann. N. Y. Acad. Sci.* **2011**, *1223*, 69–81. [CrossRef] [PubMed]
8. Caro, T.M.; O'Doherty, G. On the Use of Surrogate Species in Conservation Biology. *Conserv. Biol.* **1999**, *13*, 805–814. [CrossRef]
9. Sergio, F.; Caro, T.; Brown, D.; Clucas, B.; Hunter, J.; Ketchum, J.; McHugh, K.; Hiraldo, F. Top Predators as Conservation Tools: Ecological Rationale, Assumptions, and Efficacy. *Annu. Rev. Ecol. Evol. Syst.* **2008**, *39*, 1–19. [CrossRef]
10. Natsukawa, H.; Sergio, F. Top Predators as Biodiversity Indicators: A Meta-Analysis. *Ecol. Lett.* **2022**, *25*, 2062–2075. [CrossRef]
11. Sergio, F.; Newton, I.; Marchesi, L. Conservation: Top Predators and Biodiversity. *Nature* **2005**, *436*, 192. [CrossRef]
12. Ozaki, K.; Isono, M.; Kawahara, T.; Iida, S.; Kudo, T.; Fukuyama, K. A Mechanistic Approach to Evaluation of Umbrella Species as Conservation Surrogates. *Conserv. Biol.* **2006**, *20*, 1507–1515. [CrossRef] [PubMed]

13. Burgas, D.; Byholm, P.; Parkkima, T. Raptors as Surrogates of Biodiversity along a Landscape Gradient. *J. Appl. Ecol.* **2014**, *51*, 786–794. [CrossRef]
14. Senzaki, M.; Yamaura, Y.; Nakamura, F. The Usefulness of Top Predators as Biodiversity Surrogates Indicated by the Relationship between the Reproductive Outputs of Raptors and Other Bird Species. *Biol. Conserv.* **2015**, *191*, 460–468. [CrossRef]
15. Kenward, R. *The Goshawk*; T & AD Poyser: Berkhamsted, UK, 2006; ISBN 978-0-7136-6565-9.
16. Mirski, P.; Grosberg, J.; Kull, T.; Mellov, P.; Tõnisalu, G.; Väli, V.; Väli, Ü. Movement of avian predators points to biodiversity hotspots in agricultural landscape. *Ecol. Ind.*. in press.
17. Natsukawa, H. Raptor Breeding Sites as a Surrogate for Conserving High Avian Taxonomic Richness and Functional Diversity in Urban Ecosystems. *Ecol. Indic.* **2020**, *119*, 106874. [CrossRef]
18. Natsukawa, H. Raptor Breeding Sites Indicate High Taxonomic and Functional Diversities of Wintering Birds in Urban Ecosystems. *Urban For. Urban Green.* **2021**, *60*, 127066. [CrossRef]
19. Jokimäki, J.; Suhonen, J.; Inki, K.; Jokinen, S. Biogeographical Comparison of Winter Bird Assemblages in Urban Environments in Finland. *J. Biogeogr.* **1996**, *23*, 379–386. [CrossRef]
20. Tornberg, R.; Korpimäki, E.; Byholm, P. Ecology of the Northern Goshawk in Fennoscandia. *Stud. Avian Biol.* **2006**, *31*, 141–157.
21. Järvinen, O.; Väisanen, R. Quantitative Biogeography of Finnish Land Birds as Compared with Regionality in Other Taxa. *Ann. Zool. Fenn.* **1980**, *17*, 67–85.
22. Goławski, A.; Dombrowski, A. The Effects of Weather Conditions on the Numbers of Wintering Birds and the Diversity of Their Assemblages in Villages and Crop Fields in East-Central Poland. *Ital. J. Zool.* **2011**, *78*, 364–369. [CrossRef]
23. Tryjanowski, P.; Sparks, T.H.; Biaduń, W.; Brauze, T.; Hetmański, T.; Martyka, R.; Skórka, P.; Indykiewicz, P.; Myczko, Ł.; Kunysz, P.; et al. Winter Bird Assemblages in Rural and Urban Environments: A National Survey. *PLoS ONE* **2015**, *10*, e0130299. [CrossRef]
24. Zimaroyeva, A.A.; Matsyura, A.V.; Jankowsky, K. Spatial Patterns of Habitat Distribution of Corvidae (the Case of Urban-Rural Gradient). *Biosyst. Divers.* **2016**, *24*, 451–458. [CrossRef]
25. Jeltsch, F.; Bonte, D.; Pe'er, G.; Reineking, B.; Leimgruber, P.; Balkenhol, N.; Schröder, B.; Buchmann, C.M.; Mueller, T.; Blaum, N.; et al. Integrating Movement Ecology with Biodiversity Research—Exploring New Avenues to Address Spatiotemporal Biodiversity Dynamics. *Mov. Ecol.* **2013**, *1*, 6. [CrossRef] [PubMed]
26. Dickinson, J.L.; Shirk, J.; Bonter, D.; Bonney, R.; Crain, R.L.; Martin, J.; Phillips, T.; Purcell, K. The Current State of Citizen Science as a Tool for Ecological Research and Public Engagement. *Front. Ecol. Environ.* **2012**, *10*, 291–297. [CrossRef] [PubMed]
27. Bonney, R.; Shirk, J.L.; Phillips, T.B.; Wiggins, A.; Ballard, H.L.; Miller-Rushing, A.J.; Parrish, J.K. Next Steps for Citizen Science. *Science* **2014**, *343*, 1436–1437. [CrossRef] [PubMed]
28. Fraisl, D.; Hager, G.; Bedessem, B.; Gold, M.; Hsing, P.-Y.; Danielsen, F.; Hitchcock, C.B.; Hulbert, J.M.; Piera, J.; Spiers, H.; et al. Citizen Science in Environmental and Ecological Sciences. *Nat. Rev. Methods Prim.* **2022**, *2*, 64. [CrossRef]
29. Dickinson, J.L.; Zuckerberg, B.; Bonter, D.N. Citizen Science as an Ecological Research Tool: Challenges and Benefits. *Annu. Rev. Ecol. Evol. Syst.* **2010**, *41*, 149–172. [CrossRef]
30. Chandler, M.; See, L.; Copas, K.; Bonde, A.M.Z.; López, B.C.; Danielsen, F.; Legind, J.K.; Masinde, S.; Miller-Rushing, A.J.; Newman, G.; et al. Contribution of Citizen Science towards International Biodiversity Monitoring. *Biol. Conserv.* **2017**, *213*, 280–294. [CrossRef]
31. Pocock, M.J.O.; Chandler, M.; Bonney, R.; Thornhill, I.; Albin, A.; August, T.; Bachman, S.; Brown, P.M.J.; Cunha, D.G.F.; Grez, A.; et al. A Vision for Global Biodiversity Monitoring With Citizen Science. *Adv. Ecol. Res.* **2018**, *59*, 169–223. [CrossRef]
32. McKinley, D.C.; Miller-Rushing, A.J.; Ballard, H.L.; Bonney, R.; Brown, H.; Cook-Patton, S.C.; Evans, D.M.; French, R.A.; Parrish, J.K.; Phillips, T.B.; et al. Citizen Science Can Improve Conservation Science, Natural Resource Management, and Environmental Protection. *Biol. Conserv.* **2017**, *208*, 15–28. [CrossRef]
33. Statistics Estonia Weather. Available online: https://andmed.stat.ee/en/stat (accessed on 23 May 2023).
34. Landboard Estonian Topographic Database. Available online: https://geoportaal.maaamet.ee/eng/Spatial-Data/Estonian-Topographic-Database-p305.html (accessed on 15 April 2023).
35. Abarenkov, K.; Tedersoo, L.; Nilsson, R.H.; Vellak, K.; Saar, I.; Veldre, V.; Parmasto, E.; Prous, M.; Aan, A.; Ots, M.; et al. Plutof-a Web Based Workbench for Ecological and Taxonomic Research, with an Online Implementation for Fungal Its Sequences. *Evol. Bioinform.* **2010**, *2010*, 189–196. [CrossRef]
36. Cabeza, M.; Arponen, A.; Van Teeffelen, A. Top Predators: Hot or Not? A Call for Systematic Assessment of Biodiversity Surrogates. *J. Appl. Ecol.* **2008**, *45*, 976–980. [CrossRef]
37. Roth, T.; Weber, D. Top Predators as Indicators for Species Richness? Prey Species Are Just as Useful. *J. Appl. Ecol.* **2008**, *45*, 987–991. [CrossRef]
38. Santangeli, A.; Girardello, M. The Representation Potential of Raptors for Globally Important Nature Conservation Areas. *Ecol. Indic.* **2021**, *124*, 107434. [CrossRef]
39. Väli, Ü.; Mirski, P.; Sein, G.; Abel, U.; Tõnisalu, G.; Sellis, U. Movement Patterns of an Avian Generalist Predator Indicate Functional Heterogeneity in Agricultural Landscape. *Landsc. Ecol.* **2020**, *35*, 1667–1681. [CrossRef]
40. Murgatroyd, M.; Photopoulou, T.; Underhill, L.G.; Bouten, W.; Amar, A. Where Eagles Soar: Fine-Resolution Tracking Reveals the Spatiotemporal Use of Differential Soaring Modes in a Large Raptor. *Ecol. Evol.* **2018**, *8*, 6788–6799. [CrossRef] [PubMed]

41. Krone, O.; Treu, G. Movement Patterns of White-Tailed Sea Eagles near Wind Turbines. *J. Wildl. Manag.* **2018**, *82*, 1367–1375. [CrossRef]
42. Mirski, P.; Väli, Ü. Movements of Birds of Prey Reveal the Importance of Tree Lines, Small Woods and Forest Edges in Agricultural Landscapes. *Landsc. Ecol.* **2021**, *36*, 1409–1421. [CrossRef]

diversity

MDPI

Article

The White-Tailed Eagle, the Apex Predator, Adjusts Diet towards Larger Prey in Suboptimal Territories

Paweł Mirski [1,*] and Ervin Komar [2]

1 Faculty of Biology, University of Białystok, Ciołkowskiego 1J, 15-245 Białystok, Poland
2 Pajūrio Tyrimų ir Planavimo Institutas, Baltijos pr. 107, 93223 Klaipėda, Lithuania; erwin.komar@gmail.com
* Correspondence: p.mirski@uwb.edu.pl

Abstract: The White-tailed eagle, an apex predator, is currently recovering its populations across Europe and has already reached high numbers in many countries. This led to the saturation of eagles in optimal habitats and their encroachment on suboptimal ones. We aimed to compare the diet of White-tailed eagles in optimal and suboptimal conditions in northeastern Poland to investigate how population development affected prey composition, which is expected to be lacking in suboptimal eagle territories. We have monitored eagle nests with trail cameras to investigate their diet objectively and precisely. In order to compare territories of different quality, we have conducted modeling of habitat suitability using data on nest locations prior to their saturation. Using recorded photos of the prey, we measured their size and estimated their weight to check if the size and biomass of the prey are comparable between optimal and suboptimal territories. We found that eagles in the latter conditions were not limited by prey biomass but turned to alternative prey and brought larger prey. The alternative prey were large birds such as White storks and Common cranes, but also chicks of other avian predators that were robbed from their nests. Most probably, eagles cope with a lack of optimal prey by ranging farther and exploring non-optimal foraging habitats. We conclude that the diet flexibility of White-tailed eagle enables him to still increase its numbers despite already high densities. Our study also shows that this species might possibly impact the White stork population, as seen in the case of the Black stork and some seabird species.

Keywords: *Haliaaetus albicilla*; alternative prey; habitat suitability; superpredation; trail cameras

Citation: Mirski, P.; Komar, E. The White-Tailed Eagle, the Apex Predator, Adjusts Diet towards Larger Prey in Suboptimal Territories. *Diversity* **2023**, *15*, 747. https://doi.org/10.3390/d15060747

Academic Editors: Vladimir D. Dobrev and Michael Wink

Received: 30 April 2023
Revised: 30 May 2023
Accepted: 5 June 2023
Published: 6 June 2023

1. Introduction

The White-tailed eagle, *Haliaeetus albicilla,* has recovered in Europe in many countries [1,2]. Locally, its numbers increased threefold in the beginning of the XXI century and are still growing [3]. In some European countries, despite the fact that it has already reached high numbers, it is still the most common species among birds of prey [4]. The increase in predator population numbers has to be supported by high availability of preferred prey or high plasticity in switching prey [5,6]. In the breeding season, White-tailed eagles are known to prey on a variety of medium and large prey, most frequently fish and waterfowl, with fish being more prevalent in inland areas and waterfowl being more prevalent in coastal areas [7–10]. In the breeding season, mammals usually form an accessory prey, but sometimes a significant one, like in the case of Scotland [11].

Given that the White-tailed eagle is an apex predator, the high increase in its density might affect its prey, but prey availability may also limit population growth and therefore force it to shift towards alternative prey. The first was mainly shown by the direct, negative impact that eagles made on seabirds: Eider *Somateria mollissima* [12] and Great cormorant *Phalacrocorax carbo* [13], but also less directly on Black stork *Ciconia nigra* in the inland [14]. The second has so far been investigated mainly by changes in diet composition during population development (increase in eagle density) [15,16]. In Lithuania, the frequency of alternative prey (i.e., terrestrial birds and mammals) and dietary niche breadth did not increase between 2005 and 2018, and authors concluded that White-tailed eagles did not rely more on alternative prey in the course of population development [16]. In Finland, between 1985 and 2010, a dietary shift was observed in which the proportion of birds increased but that of fish and mammals decreased [15]. However, revealing how the largest avian predator in Europe copes with their high food demands in the light of increasing competition and saturation of optimal habitats is still an open question. To answer it, we should quantify habitat suitability and investigate diet separately in optimal and suboptimal habitats. Heuck et al. [17] found that due to population growth, White-tailed eagles started to inhabit suboptimal habitats. In such sites, the area of suitable foraging habitat was smaller, and eagles experienced lower breeding success and a lower average number of chicks. Most probably, food availability is the link between worse foraging conditions and lower fitness in pairs in suboptimal habitats.

We have investigated the diet of White-tailed eagles with the use of trail cameras, mounted in eagles' nests to check if the prey composition differs between optimal and suboptimal habitats in the still-developing population of northeastern Poland. Territory quality was distinguished with habitat suitability modeling. We expected that in suboptimal territories, eagles will: (1) switch to alternative prey and therefore explore a wider set of prey species, compared to eagles in optimal ones, with better access to abundant and optimal-size prey; (2) compensate for a shortage of prey by hunting on larger animals; and (3) suffer from a food shortage, which will result in less food biomass brought to the nest, a lesser frequency of prey deliveries (because of a lower abundance of prey), and smaller broods. Additionally, by studying the diet of White-tailed eagles during the population expansion, we hoped to contribute to the ongoing debate on this species impact on other rare and protected species [15,18,19].

2. Materials and Methods

2.1. Study Area

The study was carried out in the Podlasie region, a part of northeastern Poland with a continental climate. The region is characterized by mosaic farmland, a few large forest complexes, and marshy river valleys. The terrain is moderately flat but shaped to some extent by glacial events. In the studied part of the region (North Podlasie Lowland), there are almost no natural lakes, but only fish ponds of small and moderate size and important river valleys of Biebrza and Narew, where high numbers of waterfowl stop at spring migration and also breed.

2.2. Habitat Suitability Models

To model habitat suitability in the studied area for the White-tailed eagle, we used data on 29 nest locations (1 nest per territory) (Figure S1) that were occupied in the studied region in the first place, up to 2010. Under the assumption that they should represent the best habitats before the population developed and saturated, if there were more nests per territory, we included only the one occupied in 2010 or the one closest to this date.

As the species is associated with river valleys, waterbodies, and large wetlands, we prepared six raster variables in QGIS 3. Four were based on a detailed vector dataset of hydrological features downloaded from the Hydrological Map of Poland dataset (Polish Geological Institute, available through https://dane.gov.pl/en, accessed 12 February 2023). The vector dataset was transformed with the "proximity" tool to get the distance to waterbodies, distance to main rivers, distance to (any) water, and distance to wetlands. Furthermore, using rasters (100 m resolution) of water and wetness (downloaded from the Copernicus Land Monitoring service), we have calculated the next two variables: the share of water and the share of wetlands (wetness) per square kilometer around the focal pixel. We used the sum of pixels around the focal one to account for the continuity of a habitat, which is important for such a large apex predator. We used the variance inflation factor (VIF) to check the collinearity between predictors and removed those with a VIF greater than 10. Four predictors remained.

Habitat suitability was modeled in R using the 'biomod2' package [20]. We have generated 10 sets of pseudoabsences with numbers three times higher than the number of occurrences. Pseudoabsences were drawn with a "disk" strategy with a minimum distance of 5000 m and a maximum of 20,000 m to presence data (actual nest locations). Models were built using the Random Forest algorithm. The data was divided into training and testing datasets in a 70:30 proportion. Next, we carried out the procedure to estimate the importance of the input variables with 10 permutations. Models were validated based on the Receiver Operation Curve (ROC). Predictors importance was estimated on a 0–1 scale with the 'biomod2' package and averaged for ten permutations.

2.3. Dietary Data

The diet White-tailed eagles was collected by trail cameras mounted in their nest at the stage of 2- to 4-week-old chicks between 2013 and 2018. Two to four trail cameras were mounted each year, about 1.5–4 m from the nest, pointing directly downwards or from a variable angle, so the nest surface was seen like a plate. In one case, the nest had fallen due to heavy wind; in another, a late chick died without obvious reasons; the trail camera was pointing above the nest; or the memory card failed. Excluding such cases, where material was incomplete, the final dataset covered 12 different nests, recorded at medium and large chick stages as well as after fledgling. We used Ltl Acorn 6210MC trail cameras with 32 GB memory cards and twelve 2500 mAh accumulators, which enabled us to follow the nest from May until October and gather at least over twelve thousand 5 MP resolution photos without replacing the accumulators. Trail cameras were set to take two photos after recording movement during the daytime, and afterwards they went into sleep mode for the next 3 min. Additionally, in the same nests, we have noted prey items according to prey remains found in and under the nests in 2011–2018. However, this was a side task, and we noted the visible prey items but did not count prey based on fish scales, fur, or other small remnants.

The territories of White-tailed eagles monitored with trail cameras were attributed to optimal and suboptimal habitats based on the cut-off value of the Random Forest habitat suitability model that minimized the absolute difference between model sensitivity and specificity. The habitat suitability value under each nest was compared with the cut-off value, and if it was lower, the territory was qualified as suboptimal and optimal if higher.

Prey was identified mostly according to expert knowledge. Most birds and mammals were identified by the authors; some consulted with other ornithologists when needed. Fish were also identified by the authors and checked with experienced anglers (mentioned in the Acknowledgments). Fish biomass was estimated from images by measuring every individual fish's total length (scale based on bird rings from photos) and comparing it with empirical data describing it [21–27]. For unidentified species, we assigned a weight of 250 g for a small fish and 400 g for a medium-sized one. Bird and mammal biomasses were estimated according to reported adult average weights and estimated weights (upon growth curves) in the case of juveniles [26–32]. The average size of unidentified pieces of meat was estimated, representing the mean prey biomass from different prey groups (320 g). Apart from the prey items themselves, the number of prey deliveries per day was noted.

2.4. Data Analysis

The impact of habitat suitability on the diet of White-tailed eagles was tested between territories in optimal and suboptimal habitats. The value of the random forest model of habitat suitability for two habitat suitability groups was tested in R software with the Wilcoxon test to check if the division into optimal and suboptimal habitats is visible and statistically supported. Brood size and diet characteristics were also compared with the Wilcoxon test between two habitat suitability groups. In tested prey groups, we have distinguished other avian predators (birds of prey and owls) to test for superpredation intensity between optimal and suboptimal habitats. Secondly, we distinguished large prey, over 3000 g, therefore beyond the size of fish and waterfowl, that White-tailed eagles are known to feed on regularly.

Prey size (the logarithm of prey weight) was investigated with linear models to explain how it was influenced by territory quality (suboptimal vs. optimal) and brood size (a numerical variable). All analyses were carried out in R 4.1.1.

3. Results

3.1. Habitat Suitability Models

Model ensembles from all Random Forest models yielded high performance (ROC = 0.958). Model sensitivity reached 100%, while specificity reached 82.4%. The cut-off value for the model was estimated to be 0.345 for habitat suitability. This threshold was used to divide the optimal (n = 6) and suboptimal (n = 6) territories of the studied species (Figure 1).

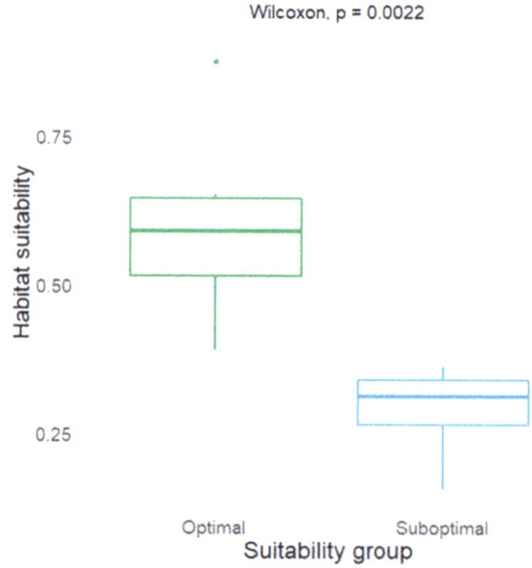

Figure 1. Comparison of the Random Forest model of habitat suitability values attributed to optimal and suboptimal territories of White-tailed eagles in northeastern Poland.

All variables contributed significantly to both models (Table 1). The distance to a waterbody was the best predictor of habitat suitability for the White-tailed eagle, followed by the distance to main rivers and to wetlands (Table 1). The sum of wetlands was still an important predictor but contributed the least.

Table 1. Habitat suitability predictors and their mean importance in Random Forest models (on a 0–1 scale, averaged for ten models) of White-tailed eagle habitats in northeastern Poland.

Model Predictors:	Mean Variable Importance	SD
distance to the main river	0.2258	0.071
distance to the waterbody	0.4014	0.107
distance to the wetland	0.2084	0.056
sum of wetlands in a 5 km radius	0.1118	0.061

3.2. Diet Composition

White-tailed eagles in northeastern Poland feed on variable prey, dominated by fish and waterfowl (Table 2). Fish dominated visibly in prey number (74%) and slightly in prey biomass (47%), followed by birds. The latter formed 21% of prey numbers but as much as

41% of prey biomass. Mammals were rarely brought (5%) to the nest, but in larger pieces, and thus contributed 12% of prey biomass.

Table 2. Diet composition of White-tailed eagles in northeastern Poland from 2013 to 2018, investigated at 12 nests using trail cameras.

Group	Species	Number	Prey Biomass (g)	Average Species Mass (g)	Share in Number	Share in Biomass	Share in Number by Habitat Optimal	Share in Number by Habitat Suboptimal
Fish	*Cyprinus carpio*	286	140,925	493	38.18%	28.98%	26.8%	46.3%
	Esoxlucius	67	22,057	329	8.95%	4.54%	13.4%	5.8%
	Cyprinidae sp.	47	18,067	384	6.28%	3.72%	5.7%	6.7%
	Tinca tinca	36	10,756	299	4.81%	2.21%	6.7%	3.5%
	Hypophthalmichthys nobilis	14	5386	385	1.87%	1.11%	2.5%	1.4%
	Abramis brama	16	5095	318	2.14%	1.05%	1.9%	2.3%
	Rutilus rutilus	20	3732	187	2.67%	0.77%	5.7%	0.5%
	Ctenopharyngodon idella	5	3300	660	0.67%	0.68%	0%	1.2%
	Carassius carassius	9	3268	363	1.20%	0.67%	1.0%	1.4%
	Lota lota	3	1540	513	0.40%	0.32%	0.6%	0.2%
	Blicca bjoerkna	6	939	157	0.80%	0.19%	0.6%	0.9%
	Perca fluviatilis	7	860	123	0.93%	0.18%	0.6%	1.2%
	Silurus glanis	2	804	402	0.27%	0.17%	0.6%	0.0%
	Oncorhynchus mykiss	2	360	180	0.27%	0.07%	0.0%	0.5%
	Salmonidae	1	125	125	0.13%	0.03%	0.0%	0.2%
	Unidentified fish	35	10,750	307	4.67%	2.21%	2.9%	6.0%
	Group total	**556**	**227,964**	**326**	**74.2%**	**46.9%**	**69.0%**	**78.1%**
Birds	*Ciconia Ciconia (adultus)*	22	87,550	3980	2.94%	18.01%	1.6%	3.9%
	Fulica atra	34	24,430	719	4.54%	5.02%	6.1%	3.5%
	Anser anser (pullus)	30	22,410	747	4.01%	4.61%	8.0%	1.2%
	Grus grus (adultus)	4	20,600	5150	0.53%	4.24%	0.6%	0.5%
	Gallus gallus domesticus	7	19,600	2800	0.93%	4.03%	1.0%	0.9%
	Anas platyrhynchos	7	6700	957	0.93%	1.38%	1.3%	0.7%
	Buteo buteo	13	5850	450	1.74%	1.20%	0.0%	3.0%
	Columbasp.	4	1440	360	0.53%	0.30%	0.6%	0.5%
	Corvus cornix	4	1190	298	0.53%	0.24%	1.0%	0.2%
	Corvus corax	1	1100	1100	0.13%	0.23%	0.3%	0.0%
	Chroicocephalus ridibundus	6	780	130	0.80%	0.16%	1.9%	0.0%
	Chlidonias sp.	6	610	102	0.80%	0.13%	1.9%	0.0%
	Cygnus sp. (pullus)	1	600	600	0.13%	0.12%	0.3%	0.0%
	Asio otus	3	445	148	0.40%	0.09%	0.0%	0.7%
	Strix aluco	1	350	350	0.13%	0.07%	0.0%	0.2%
	Tachybaptus ruficollis	1	170	170	0.13%	0.03%	0.0%	0.2%
	Sturnus vulgaris	1	80	80	0.13%	0.02%	0.0%	0.2%
	Turdus sp.	1	70	70	0.13%	0.01%	0.0%	0.2%
	Unidentified birds	15	5900	393	2.00%	1.21%	3.2%	1.2%
	Group total	**161**	**199,875**	**979**	**21.5%**	**41.1%**	**27.8%**	**17.1%**

Table 2. *Cont.*

Group	Species	Number	Prey Biomass (g)	Average Species Mass (g)	Share in Number	Share in Biomass	Share in Number by Habitat Optimal	Share in Number by Habitat Suboptimal
Mammals	*Lepus europaeus*	4	13,000	3250	0.53%	2.67%	0.0%	0.9%
	Castor fiber	2	12,000	6000	0.27%	2.47%	0.0%	0.2%
	Capreolus capreolus	5	11,700	2340	0.67%	2.41%	1.0%	0.5%
	Canis lupus familiaris	1	4000	4000	0.13%	0.82%	0.3%	0.0%
	Sus scrofa/domesticus	1	2500	2500	0.13%	0.51%	0.0%	0.2%
	Martes martes	1	1100	1100	0.13%	0.23%	0.0%	0.2%
	Sus scrofa	1	1000	1000	0.13%	0.21%	0.0%	0.2%
	Unidentified mammals	17	13,110	771	2.27%	2.70%	1.9%	2.5%
	Group total	**32**	**58,410**	**2620**	**4.3%**	**12.01%**	**3.2%**	**4.7%**
	TOTAL	**749**	**486,249**	-	**100%**	**100%**	**100%**	**100%**

We found 15 species of fish, representing both artificial fish ponds and natural rivers. The carp, *Cyprinus carpio*, constituted 55% of all the identified fish species brought to the nest. The average mass of this prey species reached 493 g, with a maximum of 2120 g. Other fish species averaged 439 g, ranging from 71 g to 1100 g. The share of carp in prey numbers and biomass shows the importance of the fish ponds for the species in this lakeless part of the region (north Podlasie Lowland).

White-tailed eagles fed on at least seventeen species of birds and a domestic chicken (Table 2). Waterfowl (ducks, geese, rails, swans, and grebes) constituted exactly half of the identified species. Other waterbirds, such as gulls and terns, added another 7% to the bird prey number. Important alternative prey groups were the large birds, such as White stork *Ciconia ciconia* and Common crane *Grus grus* (18% of identified birds), but also chicks of other avian predators that must have been taken from their nests, mostly Common buzzard, but also Long-eared owl (11% of bird prey number). Single adults or juveniles of Common ravens and Tawny owls were also noted.

Mammals were the least frequent but locally important in terms of biomass. Young beavers were probably hunted in the water, but other species were possibly taken as carcasses, some of which were likely roadkill. In many cases, we were not able to identify the mammal species as only a piece of meat with fur was recorded at the nest.

Prey remains found in the nest show a more or less similar variety of prey, but with a lesser diversity of fish and a greater diversity of bird species, especially the waterfowl (Table S1). No signs of superpredation were noted in prey remains. This is because White-tailed eagles predate mostly on chicks, yet without contour feathers, they are eaten entirely and leave no trace. The ratio of main prey groups based on nest remains is completely different (Table S2). Birds dominated (80%) in prey remains, while in diets recorded by trail cameras, they were almost four times less frequent (21%). On the contrary, fish (which are eaten whole) were found only rarely in prey remains (14%), but were dominant in the data investigated from trail cameras (74%). Mammals did not differ much between those two datasets.

3.3. Factors Affecting the White-Tailed Eagle Diet

Proportions of main prey groups did not differ significantly between eagles in optimal and suboptimal territories (Table S3), although the difference in the ratio of mammals was slightly higher in optimal territories (Figure S2) but only close to significant (p = 0.076). Detailed comparisons showed eagles fed much more frequently on carp but much less on Northern pike (*Esox lucius*) and Common roach (*Rutilus rutilus*) in suboptimal than optimal territories. In the case of bird prey, eagles preyed on white storks and other avian predators much more frequently in suboptimal territories but rarely (compared to optimal ones) on Eurasioan coots, *Fulica atra*, and Greylag geese, *Anser Anser*.

To some extent, White-tailed eagles altered their prey preferences depending on habitat quality. In optimal habitats, they hardly fed on other predators, while they did so relatively frequently in suboptimal habitats (Figure 2a). The ratio of superpredation was significantly higher in prey number and prey biomass in suboptimal habitats than optimal habitats (Table S3). Furthermore, we noticed that eagles in suboptimal habitats often brought large prey, such as storks, cranes, and mammals, to their nests (Figure 2b). This result was not statistically significant, but taking into account the small sample size of nests, we consider it to be close to significant. Lastly, we did not find any difference in prey species richness between optimal and suboptimal territories (Table S3).

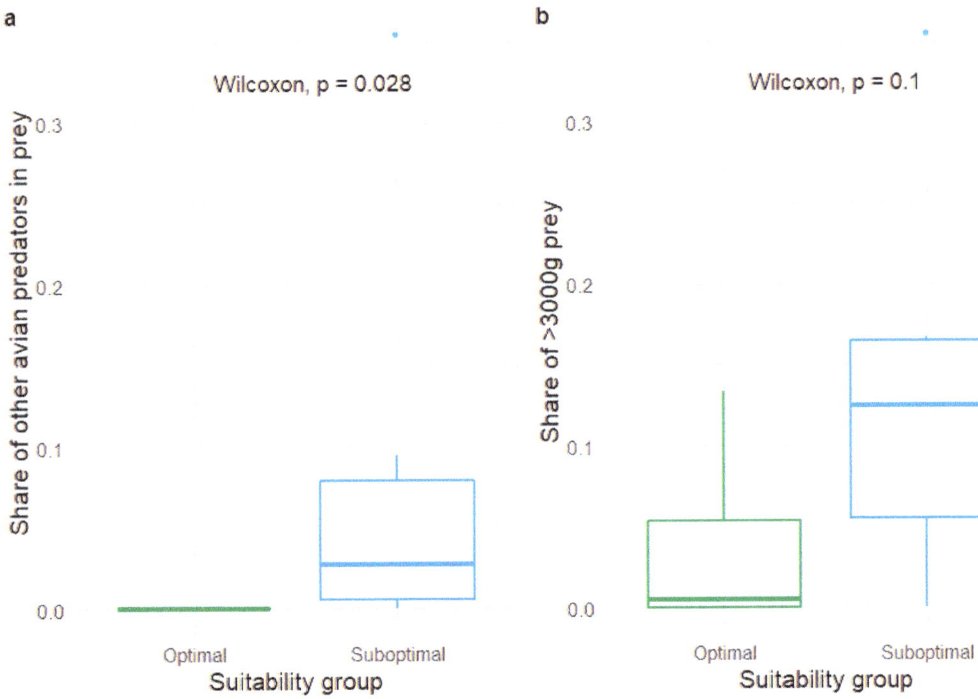

Figure 2. Comparison of the share of the two alternative prey groups: other raptors (**a**) and large prey (**b**), in the diet of the White-tailed eagles occupying optimal and suboptimal territories in northeastern Poland.

3.4. Prey Size

We tested the effects of territory quality and brood size on prey size, prey biomass, and the number of prey deliveries. Brood size did not differ between optimal and suboptimal habitats (p = 0.93), although the only brood of three chicks that successfully fledged was recorded in an optimal territory. We did not find a territory quality impact on daily biomass consumption per nest or the daily number of prey deliveries per nest (Figure 3a,b, Table S3). However, we found that the total biomass brought to the nest was greater in broods with multiple chicks (Figure 3c). The number of prey deliveries seemed slightly lower in solitary broods, but not significantly (Figure 3d).

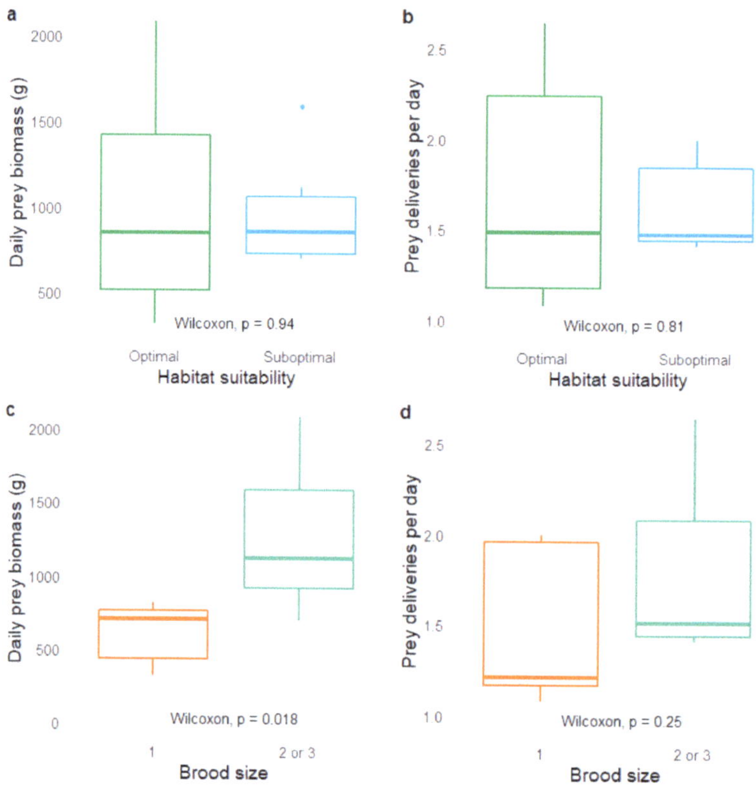

Figure 3. Daily prey biomass and the number of prey deliveries per day, recorded in the nests of White-tailed eagles in northeastern Poland and compared between optimal and suboptimal territories (**a**,**b**) and broods with single or multiple chicks (**c**,**d**).

White-tailed eagles fed on larger birds and fish in suboptimal habitats (Table 3). In the case of mammals, their sizes were smaller in suboptimal habitats, but due to the small sample size, the results were not significant (Figure S3). When raising larger broods, eagles selected larger fish, but the same relationship was not significant in the case of bird prey (Table 3). The effect of territory quality was greater than that of brood size. The models explained almost 10% of the variance in fish size, but only almost 5% in the case of bird prey. Overall, in optimal territories, eagles most often brought prey weighing 500–1000 g to their nest.

Table 3. Linear models explaining single White-tailed eagles' prey item mass with territory quality (suboptimal/optimal) and the brood size.

	Bird Prey Mass			Fish Prey Mass		
Predictors	**Estimates**	**CI**	***p***	**Estimates**	**CI**	***p***
(Intercept)	6.17	5.60–6.74	<0.001	5.44	5.25–5.63	<0.001
Territory (suboptimal)	0.45	0.07–0.84	0.020	0.29	0.17–0.41	<0.001
Brood size	0.15	−0.09–0.39	0.216	0.18	0.09–0.27	<0.001
Observations		131			340	
R^2/R^2 adjusted		0.043/0.028			0.094/0.089	

4. Discussion

4.1. Diet Differences between Optimal and Suboptimal Habitats

We found that the diet of White-tailed eagles was affected by territory quality, but not all of our predictions were confirmed. Eagles switched to alternative prey, but to a limited extent. Dietary breadth, measured by the number of species that were brought to the nest, did not differ with regard to territory quality. However, eagles reached for alternative prey such as large birds other than waterfowl and robbed the nests of other avian predators, mainly the Common buzzard. The latter was already reported in Lithuania [19], but here we were able to show this prey was clearly more frequently used in suboptimal territories. Secondly, we confirmed that in suboptimal conditions in comparison to optimal ones, eagles will compensate for abundant prey with larger fish and birds (but not mammals). In optimal habitats, the most frequent prey ranged from 500 to 1000 g, but it was much more variable in suboptimal habitats and included a relatively high share of prey over 3000 g. The dominant prey size in optimal habitats corresponded to the ones recorded at Estonian fishponds (optimal foraging habitat), where White-tailed eagles preyed on 300–1050 g fish [33]. Interestingly, in the lakeland adjacent to our study, about 20 years before, eagles were estimated to prey on 514 g of fish and 611 g of birds on average [34]. This confirms that in good-quality habitats (such as large lakes) and before the population saturated, White-tailed eagles foraged on relatively small prey, that was possibly abundant and easy to handle and carry.

Additionally, we found that bigger brood also contributed to selection for larger fish, but to a lesser extent than territory quality. In broods of twins or triplets, the prey biomass was greater than in nests with single chicks, but not the number of prey deliveries, which confirms that eagles compensated for higher food demands with larger prey. Finally, we could not confirm that in suboptimal territories, eagles would suffer from food shortages. The daily biomass and number of prey deliveries, as well as the number of chicks per brood, were similar in optimal and suboptimal habitats.

4.2. Mitigation of Lower-Quality Territories with Larger Prey

In Greece, White-tailed eagles occupying territories of potentially similar quality at neighboring lakes were found to hunt on similar prey species of similar body mass. It suggests that in comparable conditions, eagles select locally optimal prey, also in terms of their size [35]. In our case, where territories differed in their quality, we observed a clear difference in selected prey size. In suboptimal territories, eagles brought larger prey to the nest, while in optimal habitats, the most frequent size of prey was much smaller. It seems that the first had to spend additional energy to find larger prey, handle it, and transport it to the nest. Most likely, it required a much larger area to search for bigger prey, but eagles also had to explore habitats that were different from optimal ones (waterbodies, river valleys). The latter was seen as relatively high numbers of adult white storks, which were possibly hunted over agricultural landscapes and by robbing the nests of other avian predators, which requires penetration of the forest. To prove that White-tailed eagles have to range further to find sufficient food for their offspring, we would need to track them with GPS telemetry devices. We did not have such data for the eagles whose nests were monitored by us with trail cameras, but we had followed another adult male, who occupied suboptimal territory on the edges of Biebrza Valley but had no access to the best foraging sites near the river. During most of the season, he ranged over a small area, but when having chicks, the male flew regularly 25 km to the large waterbody (Mirski, unpublished). A similar case was observed in Lesser spotted eagles, *Clanga pomarina*, which were also forced to forage exceptionally far from their nest to successfully raise chicks in a situation of insufficient foraging areas [36].

A mechanistic population model showed that the increase in density of White-tailed eagles should affect their breeding performance [15]. However, empirical data from Lithuania pointed out that eagles were not food-limited, nor did their reproduction decrease, despite a high increase in density [3,14]. Compensation of poor territory quality by the increase of home ranges to hunt for larger prey might be the best explanation for why the theoretical model and empirical data did not match in this case.

4.3. The Impact of a Developing Apex Predator Population on Its Prey

The growing population of an apex predator often raises questions about its impact on prey species, both livestock and wild animals. Regarding the first, White-tailed eagles pose only a small or moderate threat [8]. Regarding the second, for some prey species, the pressure from this predator can be significant. Thus far, this has mainly been proven for seabirds such as the Common eider [10] and the Black-legged kittiwake *Rissa tridactyla* [37], which are top-down controlled by the eagles. It is difficult to assess, though, if the current impact is different from a century ago, before the numbers of this species dropped and recovered again.

There is less proof on the impact of the White-tailed eagle on its prey in inland areas, but a recent study shows that they can rob nests of other predators and the Black stork ([14,17,38], this study, and our own unpublished data). Not only by nest robbing but possibly also by killing adult storks, White-tailed eagles can affect other species, particularly Black storks, by creating a "landscape of fear", in which storks avoid nesting within a few kilometers of an eagle's nest. This affects nest occupancy rates and breeding performance in this declining species. Our study shows evidence that the White stork is also directly threatened by White-tailed eagles. We found that 13% of birds brought to the nest were adult storks, some even carrying rings, proving they were mature birds. For a long-lived species like storks, a high surplus mortality rate from predation can be reflected in a significant population decline. Similarly, in the case of predated Common cranes, which were, however, found less often in our study, an even higher ratio of White storks and Common cranes to other prey was found in our opportunistically collected data on prey remains in eagles' nests (Table S2). However, this dataset is biased towards durable leftovers and can lead to the "demonization" of this apex predator, which, in fact, poses some threat to particular prey species but, in optimal territories, relies mostly on fish and waterfowl.

Supplementary Materials: The following supporting information can be downloaded at: https://www.mdpi.com/article/10.3390/d15060747/s1. Table S1. White-tailed eagle diet investigated from prey remains found in different nests or seasons in northeastern Poland in 2011–2018; Table S2. Differences in general prey composition of White-tailed eagles from northeastern Poland, studied by examination of prey remains in their nests (2011–2018) and identification of prey from the photos taken by trail cameras recording nests (2013–2018); Table S3. Results of statistical comparisons between diet characteristics of White-tailed eagles in optimal (n = 6) and suboptimal (n = 6) habitats in northeastern Poland, studied with trail cameras at the nest and tested with the Wilcoxon test; Figure S1. Habitat suitability (Random Forest model) for the White-tailed eagle in northeastern Poland; Figure S2. Proportions of main prey groups in the White-tailed eagle diet in territories of optimal and suboptimal habitats; Figure S3. Prey size of the main prey groups of White-tailed eagles, measured from trail camera images recording their nests in optimal and suboptimal habitats in northeastern Poland; Figure S4. Histograms of 484 prey-size items from White-tailed eagle nests in northeastern Poland, identified and measured from trail camera images in the nests located in optimal and suboptimal habitats.

Author Contributions: Conceptualization, P.M.; methodology, P.M. and E.K.; investigation, P.M. and E.K.; data curation, P.M. and E.K.; writing—original draft preparation, P.M.; writing—review and editing, P.M. and E.K.; visualization, P.M. All authors have read and agreed to the published version of the manuscript.

Funding: This research was funded by the National Geographic Society, grant number GEFNE75-13.

Institutional Review Board Statement: The license to enter White-tailed eagle nests and monitor them with trail cameras was granted by the Regional Inspectorate for Environmental Protection in Białystok.

Data Availability Statement: Key data are given in Table 2 and Table S1. Additional data is available upon request.

Acknowledgments: We would like to thank Krzysztof Henel for help in collecting data in Biebrza National Park, Marcin Leszczyński, Tomasz Tumiel, and Grzegorz Grygoruk for help in the fieldwork, Sylwester Aftyka for help in the identification of bird feathers, Marcin Wereszczuk, and Ryszard Szerakowski for their help in fish identification.

Conflicts of Interest: The authors declare no conflict of interest.

References

1. Dementavičius, D. White-tailed eagle (*Haliaeetus albicilla*) in Lithuania: Population numbers and trends 1900–2007. *Acta Zool.* **2007**, *17*, 281–285. [CrossRef]
2. Sulawa, J.; Robert, A.; Köppen, U.; Hauff, P.; Krone, O. Recovery dynamics and viability of the white-tailed eagle (*Haliaeetus albicilla*) in Germany. *Biodivers. Conserv.* **2010**, *19*, 97–112. [CrossRef]
3. Dementavičius, D.; Rumbutis, S.; Treinys, R. Spatial and temporal variation in breeding performance in the increasing White-tailed Eagle *Haliaeetus albicilla* population to the east of the Baltic Sea. *Bird Study* **2022**, *68*, 462–476. [CrossRef]
4. Wardecki, Ł.; Chodkiewicz, T.; Beuch, S.; Smyk, B.; Sikora, A.; Neubauer, G.; Meissner, W.; Marchowski, D.; Wylegała, P.; Chylarecki, P. Monitoring Ptaków Polski w latach 2018–2021. *Biul. Monit. Przyrody* **2021**, *22*, 1–80.
5. Galushin, V.M. Synchronous fluctuations in populations of some raptors and their prey. *IBIS* **1974**, *116*, 127–134. [CrossRef]
6. Newton, I. *Population Ecology of Raptors*; T & AD Poyser: Berkhamsted, UK, 1979.
7. Sulkava, S.; Tornberg, R.; Koivusaari, J. Diet of the white-tailed eagle *Haliaeetus albicilla* in Finland. *Ornis Fenn.* **1997**, *74*, 65–78.
8. Nadjafzadeh, M.; Hofer, H.; Krone, O. The link between feeding ecology and lead poisoning in white-tailed eagles. *J. Wildl. Manag.* **2013**, *77*, 48–57. [CrossRef]
9. Sándor, A.; Alexe, V.; Marinov, M.; Doroşencu, A.; Domşa, C.; Kiss, B.J. Nest-site selection, breeding success, and diet of white-tailed eagles (*Haliaeetus albicilla*) in the Danube Delta, Romania. *Turk. J. Zool.* **2015**, *39*, 300–307. [CrossRef]
10. Ekblad, C.; Tikkanen, H.; Sulkava, S.; Laaksonen, T. Diet and breeding habitat preferences of White-tailed Eagles in a northern inland environment. *Polar Biol.* **2020**, *43*, 2071–2084. [CrossRef]
11. Whitfield, D.P.; Marquiss, M.; Reid, R.; Grant, J.; Tingay, R.; Evans, R.J. Breeding season diets of sympatric White-tailed Eagles and Golden Eagles in Scotland: No evidence for competitive effects. *Bird Study* **2013**, *60*, 67–76. [CrossRef]
12. Morelli, F.; Laursen, K.; Svitok, M.; Benedetti, Y.; Møller, A.P. Eiders, nutrients and eagles: Bottom-up and top-down population dynamics in a marine bird. *J. Anim. Ecol.* **2021**, *90*, 1844–1853. [CrossRef]
13. Bregnballe, T.; Tofft, J.; Kotzerka, J.; Lehikoinen, A.; Rusanen, P.; Herrmann, C.; Krone, O.; Engström, H.; Rattiste, K.; Reich, J.; et al. Occurrence and behaviour of White-tailed Eagles *Haliaeetus albicilla* in Great Cormorant *Phalacrocorax carbo sinensis* colonies in countries around the Baltic Sea. *Ardea* **2021**, *109*, 565–582. [CrossRef]
14. Zawadzki, G.; Zawadzki, J.; Drozdowski, S.; Zawadzka, D. The avoidance of living in the vicinity of a top predator: The coexistence of the black stork and the white-tailed eagle in NE Poland. *Eur. Zool. J.* **2022**, *89*, 1223–1237. [CrossRef]
15. Ekblad, C.; Sulkava, S.; Stjernberg, T.; Laaksonen, T. Landscape-Scale Gradients and Temporal Changes in the Prey Species of the White-Tailed Eagle (*Haliaeetus albicilla*). *Ann. Zool. Fenn.* **2016**, *53*, 228–240. [CrossRef]
16. Dementavičius, D.; Rumbutis, S.; Virbickas, T.; Vaitkuvienė, D.; Dagys, M.; Treinys, R. Spatial and temporal variations in the White-tailed Eagle *Haliaeetus albicilla* breeding diet revealed by prey remains. *Bird Study* **2020**, *67*, 206–216. [CrossRef]
17. Heuck, C.; Herrmann, C.; Schabo, D.G.; Brandl, R.; Albrecht, J. Density-dependent effects on reproductive performance in a recovering population of White-tailed Eagles *Haliaeetus albicilla*. *IBIS* **2017**, *159*, 297–310. [CrossRef]
18. Treinys, R.; Dementavičius, D.; Mozgeris, G.; Skuja, S.; Saulius, R.; Stončius, D. Coexistence of protected avian predators: Does a recovering population of White-tailed Eagle threaten to exclude other avian predators? *Eur. J. Wildl. Res.* **2021**, *57*, 1165–1174. [CrossRef]
19. Kamarauskaitė, A.; Dementavičius, D.; Skuja, S.; Dagys, M.; Treinys, R. Interaction between the White-tailed Eagle and Common Buzzard estimated by diet analysis and brood defence behaviour. *Ornis Fenn.* **2020**, *97*, 26–37.
20. Thuiller, W.; Georges, D.; Gueguen, M.; Engler, R.; Breiner, F.; Lafourcade, B.; Patin, R. Biomod2: Ensemble Platform for Species Distribution Modeling. R Package Version 4.2-3. Available online: https://cran.r-project.org/web/packages/biomod2/biomod2.pdf (accessed on 11 April 2023).
21. De Giosa, M.; Czerniejewski, P. A generalized, nonlinear regression approach to the length-weight relationship of European perch (*Perca fluviatilis* L.) from the Polish coast of the southern Baltic Sea. *Arch. Pol. Fish.* **2016**, *24*, 169–175. [CrossRef]
22. Kurbanov, A.; Bakhtiyar, K. Age and growth of bighead carp (*Hypophthalmichthys nobilis* R.) in Tudakul reservoir, Uzbekistan. *Turk. J. Fish. Aqua Sci.* **2015**, *3*, 229–232.
23. Ömer, S.; Göktuğ, G.; Mehmet, Y.; Ali, G. Some Biological Features of *Tinca tinca* Population in Asartepe Reservoir (Ankara). *J. Agric. Sci. Technol. B* **2018**, *8*, 466–471. [CrossRef]
24. Epler, P.; Popek, W.; Łuszczek-Trojnar, E.; Drąg-Kozak, E.; Szczerbik, P.; Socha, M. Age and growth rate of the roach (*Rutilus rutilus* L.) from the Solina and the Tresna (Zywieckie Lake) dam reservoirs. *Acta Sci. Pol. Piscaria* **2005**, *4*, 59–70.
25. Stavrescu-Bedivan, M.; Vasile Scaeteanu, G.; Madjar, R.M.; Manole, M.S. Investigation of length-weight relationship and condition factor of *Carassius gibelio* related to water quality in Pantelimon II lake. *AgroLife Sci. J.* **2018**, *7*, 123–130.

26. Karataş, M.; Çiçek, E.; Başusta, A.; Başusta, N. Age, Growth and Mortality of Common Carp (*Cyprinus carpio* Linneaus, 1758) Population in Almus Dam Lake (Tokat-Turkey). *J. Appl. Biol. Sci.* **2007**, *1*, 81–85.
27. Yilmaz, S.; Yazicioğlu, O.; Yazici, R.; Polat, N. Age, Growth and Reproductive Period of White Bream, Blicca bjoerkna (L., 1758) in Lake Ladik, Turkey. *J. Limnol. Freshw. Fish. Res.* **2015**, *1*, 9–18. [CrossRef]
28. Jedrzejewska, B.; Jedrzejewski, W. Predation in Vertebrate Communities: The Białowieża Primeval Forest as a Case Study. *Ecol. Stud.* **1998**, *135*, 462. [CrossRef]
29. Tsachalidis, E.P.; Liordos, V.; Goutner, V. Growth of White Stork *Ciconia ciconia* nestlings. *Ardea* **2005**, *93*, 133–137.
30. Lokemoen, J.T.; Johnson, D.H.; Sharp, D.E. Weights of wild Mallard *Anas platyrhynchos*, Gadwall *A. strepera*, and Blue-winged Teal *A. discors* during the breeding season. *Wildfowl* **1990**, *41*, 122–132.
31. Meissner, W. Analiza biometryczna łysek *Fulica atra* z północnej Polski. *Ornis Pol.* **2010**, *51*, 21–29.
32. Owen, M.; Cook, W.A. Variations in body weight, wing length and condition of Mallard *Anas platyrhynchos platyrhynchos* and their relationship to environmental changes. *J. Zool.* **1977**, *183*, 377–395. [CrossRef]
33. Tuvi, J.; Väli, Ü. The impact of the White-tailed Eagle *Haliaeetus albicilla* and the Osprey Pandion haliaetus on Estonian Common Carp *Cyprinus carpio* production: How large is the economic loss? *Proc. Est. Acad. Sci. Biol./Ecol.* **2007**, *56*, 209–223. [CrossRef]
34. Zawadzka, D. Feeding habits of the Black Kite Milvus migrans, Red Kite Milvus milvus, White-tailed Eagle Haliaeetus albicilla and Lesser Spotted Eagle Aquila pomarina in Wigry National Park (NE Poland). *Acta Ornithol.* **1999**, *34*, 65–75.
35. Bounas, A.; Karta, E.; Navarrete, E.; Sidiropoulos, L.; Alivizatos, H. Diet composition of White-tailed Eagles inhabiting two adjacent inland lakes in Northern Greece. *Raptor J.* **2023**, *17*. [CrossRef]
36. Mirski, P.; Cenian, Z.; Dagys, M.; Daróczi, S.; Dementavičius, D.; Maciorowski, G.; Menderski, S.; Nowak, D.; Pongrácz, Á.; Prommer, M.; et al. Sex-, landscape- and climate-dependent patterns of home-range size—A macroscale study on an avian generalist predator. *IBIS* **2021**, *163*, 641–657. [CrossRef]
37. Anker-Nilssen, T.; Fayet, A.L.; Aarvak, T. Top-down control of a marine mesopredator: Increase in native white-tailed eagles accelerates the extinction of an endangered seabird population. *J. Appl. Ecol.* **2023**, *60*, 445–452. [CrossRef]
38. Yurko, V. Diet of the White-Tailed Eagle During the Breeding Season in the Polesski State Radiation-Ecological Reserve, Belarus. *Raptors Conserv.* **2016**, *32*, 21–31. [CrossRef]

diversity

MDPI

Article

Rabbits or Refuse? Landfill Use and Relevance as a Food Source for an Increasing Wintering Population of the Red Kite

Íñigo Vicente-Hernández [1], Félix Martínez [2] and Guillermo Blanco [1,*]

[1] Department of Evolutionary Ecology, Museo Nacional de Ciencias Naturales, CSIC, José Gutiérrez Abascal 2, 28006 Madrid, Spain; inigo.23.7@gmail.com
[2] Escuela Internacional de Doctorado, Universidad Rey Juan Carlos (URJC), C/Tulipán s/n, 28933 Móstoles, Spain; femolivas@gmail.com
* Correspondence: g.blanco@csic.es

Abstract: Household waste landfills represent a huge source of trophic resources for opportunistic and versatile wildlife species. Among them, the red kite (*Milvus milvus*) is one of the most endangered in Europe. Several studies have pointed out the importance of landfills as a source of food for this species during the wintering season, but the information on the frequency, seasonal and daily patterns of use, and age of red kites that exploit this food source is still insufficient to understand their actual role in conservation. In this study, we evaluated the patterns of use of household waste by overwintering red kites in southeastern Madrid, central Spain. The results showed the constant presence of relatively low numbers (<30 on average) and a reduced hourly inflow and outflow of red kites throughout the day and during the whole winter period in the studied landfill. A higher proportion of juveniles was found in the landfill than in the overall wintering population. Pellet analysis clearly shows that the diet of red kites is dominated by the wild rabbit (*Oryctolagus cuniculus*), which is quantitatively very relevant compared to household waste obtained from the landfill. This suggests a relatively low quantitative importance of landfills as foraging grounds for the increasing population of wintering red kites in the study area. Instead, the high regional density of wild rabbits attracts large numbers of red kites that can eventually use landfills as a non-optimal last-resort foraging option, owing to the predictability of household waste, especially for juveniles. The continuous presence of red kites in landfills likely influences an uninformed positive perception about their relevance to the conservation of the wintering population, despite risks there faced, such as collision, electrocution, and intoxication. Future research is needed to assess in depth the influence of wild rabbits in Spain on the habitat use and global population dynamics of red kites.

Keywords: conservation; *Milvus milvus*; *Oryctolagus cuniculus*; rubbish dumps; scavenger raptors; waste management; wildlife

Citation: Vicente-Hernández, Í.; Martínez, F.; Blanco, G. Rabbits or Refuse? Landfill Use and Relevance as a Food Source for an Increasing Wintering Population of the Red Kite. *Diversity* **2023**, *15*, 704. https://doi.org/10.3390/d15060704

Academic Editors: Michael Wink and Vladimir D. Dobrev

Received: 18 April 2023
Revised: 15 May 2023
Accepted: 22 May 2023
Published: 25 May 2023

1. Introduction

The growth of the human population leads to the generation of huge quantities of household waste often deposited in open landfill sites, which represents a management challenge for environmental conservation [1–3]. In Western Europe, current regulations and standards require the progressive reduction of such practices with an ultimate goal of recycling the vast majority of these residues [4,5]. These regulations are difficult to enforce in many regions due to poor planning of urban developments, increasing consumption of waste-generating products, and a lack of practices and policies that allow for recycling [1,6,7]. Due to the expected growth and concentration of the human population in urban areas, it is expected that open landfill sites will remain active for decades to come [8,9]. Waste disposed of in these sites generates pollution that can enter soils, air, and water, negatively affecting the health of human populations living in the surroundings [3,10,11] and the health and population dynamics of wildlife species foraging there [12–14].

Household waste dumps represent a huge source of trophic resources for opportunistic and versatile wildlife species, including gulls, scavenger raptors, storks, and corvids [13–15]. Multiple environmental and population factors have been highlighted in determining the impact of waste as food on population dynamics, especially the sustained growth of populations and species most dependent on this resource. However, important aspects for understanding the patterns of landfill use are largely unknown for rare species. Among these species, the red kite (*Milvus milvus*) is one of the most endangered in Europe [16].

Numerous studies have pointed out the importance of landfills as a source of food for the red kite, especially during the wintering season [17–20]. During the breeding season, these sites are also exploited by the floating population, while breeding individuals are more dependent on wild animals that are preyed upon or scavenged [21]. There is some information on the conservation problems for red kites and other avian scavengers associated with the use of livestock carcass dumps, or "vulture restaurants" [22–26]. However, the available information on the frequency, seasonal and daily patterns of use, and age of red kites that exploit the resources found in landfills [15,27–29] is still insufficient to understand their actual role in the survival of these individuals and their importance as a whole for this species' conservation status. Previous reports often assumed that wintering populations that use landfills are highly dependent on the food there found [15,30–32]. Alternatively, landfills may act as a non-optimal last-resort foraging option owing to the predictability of food resources, albeit costly and dangerous to obtain due to competition with conspecifics and other species, and due to other risks like collision, electrocution, and intoxication [13,14]. Food exploited by red kites in landfills corresponds to small offal fragments, especially cooked or uncooked remains of poultry, lamb, cow, pig, and marine fish [33–36], as has been documented for its sister species, the black kite (*Milvus migrans*), for which there is more specific information on the use of landfills and the food there obtained [37–39].

Food remains in varying degrees of decomposition available in landfills should not provide a large amount of biomass compared to the carcasses of domestic animals disposed of in livestock carcass dumps for use by threatened scavenging birds [24,40]. These food remains are also not comparable in terms of biomass and nutritional quality with the remains of wild animals generally exploited by this species, especially key prey such as voles and rabbits, and road-killed or sick, weak, and young individuals of multiple vertebrate species [22,33,34]. In addition, carcass dumps and landfills are sources of infection by pathogens and parasites from decomposing organic matter mixed with synthetic materials and other contaminants that can affect bird health [13,14,25,26]. Regardless of the quantity and quality of the resources that kites obtain from landfills, it is challenging to determine their relevance for kite populations during wintering. This is due to the continuous nomadic movement of individuals in this season, owing to multiple factors such as weather, landfill features, the number of individuals of this and other species that use these sites, and the abundance and availability of wild prey in the surroundings. These and other important aspects for understanding the patterns of resource use in landfills and the countryside by this species are largely unknown.

In this study, we evaluated the patterns of use of household waste by overwintering red kites in central Spain. Specifically, we recorded the seasonal and hourly abundance of kites exploiting this matter and the dynamic of inflow and outflow of individuals to a particular landfill throughout the day to assess the quantitative importance of these sites for the wintering population. To evaluate the role of resources obtained at the landfill versus those provided by wild or domestic animals obtained elsewhere, the diet of red kites was studied by analyzing pellets collected at a communal roost near the landfills. We also assessed the age of individuals exploiting the resources found at the landfills and compared it with the age distribution in the total population concentrated in communal roosts in the area. The abundance of other bird species was also recorded to characterize the avian community foraging in the landfill. This information is discussed in accordance with previous knowledge of the importance and risks associated with the use of landfills by wintering red kites.

2. Materials and Methods

2.1. Study Species

The red kite is mostly distributed in Europe. It is listed in Annex I of the wild birds Directive 2009/147/EC due to its negative population trend. From 2005 until 2019, the species was listed as "Near Threatened" by the IUCN. After overcoming this negative trend and maintaining a positive trend for at least 30 years, the species was reclassified in 2020 to "Least Concern" [41], largely due to successful reintroductions in the United Kingdom and recovery partially attributed to supplementary feeding in human settlements [42,43]. The red kite migrates from central and northern Europe to more southern regions to spend the winter. These movements begin in August, reaching the wintering quarters and end in late February, when individuals return to the breeding areas, with variations depending on age and other individual and environmental factors [20,44]. Spain holds one of the most abundant breeding populations and represents its main stronghold as a wintering area, with around 50,000 individuals [32]. A sharp decline in its breeding and wintering ranges during the 1990s and 2000s [21], which continues in most regions, has led to the species being listed as "Endangered" in the Spanish National Catalogue of Threatened Species (RD 139/2011, 4 February). The decline of the red kite is mainly due to increased mortality by shooting, intentional poisoning, and unintentional intoxication through pest control poisons directed at small rodents [12,17,45–47]. Electrocutions and collisions with power lines and wind turbines also represent an increasing threat [48–51].

2.2. Study Area

The study was conducted in the southeast of Madrid, central Spain. This is a peri-urban area, highly degraded by habitat alterations due to urban developments, large infrastructure, former mining operations, and intensive irrigated agriculture [52,53]. In the area, there are two large landfills for domestic waste from the city of Madrid and other surrounding municipalities. The landfill called Mancomunidad Sur is located in the municipality of Pinto (Figure 1). This landfill (hereafter called Pinto) covers an area of 90 ha, where the household waste of a human population of 1,580,619 people is deposited [54], which is equivalent to 288,324 tons per year. This enormous amount of waste represents only 10% less than what is deposited in the other landfill in the area (Valdemingómez), the largest in the province of Madrid [55], where a large solid waste incinerator is located. The area has been noted for its high levels of soil and water contamination [53], and references therein] despite the fact that its surroundings belong to a protected area (Parque Regional del Sureste) with high landscape and biodiversity values [56]. Studies on wild birds inhabiting this area have shown high levels of multiple pollutants of different origins [53,57–59]. The red kites that feed in the aforementioned landfills establish their communal roosts in the riparian forest of the Manzanares River downstream of the city of Madrid (Figure 1).

2.3. Patterns of Use of the Landfill

Censuses were carried out at the Pinto landfill to determine the number of foraging red kites. Observations were carried out from a high and dominant point that offered a wide view of the landfill, which allowed us to count the vast majority of the birds there present. Rainy, snowy, or foggy days were avoided. Surveys were carried out by visual counts of all individuals present at a given time using binoculars. Snapshot counts were taken every 30 min, from 9:00 am to 18:00 on each sampling day, beginning on 6 December 2020 and ending on 14 February 2021. Overall, 191 counts were conducted across 11 full-day surveys; two of the hourly counts in a single day were discarded because of dense fog. Data accounting for variation in the number of kites using the landfill were grouped by weeks and months to simplify the analysis. Hourly counts throughout each sampling day were grouped into three daylight periods: (1) morning, 9:00–12:00 am; (2) midday, 12:00–15:00; and (3) afternoon, 15:00–18:00.

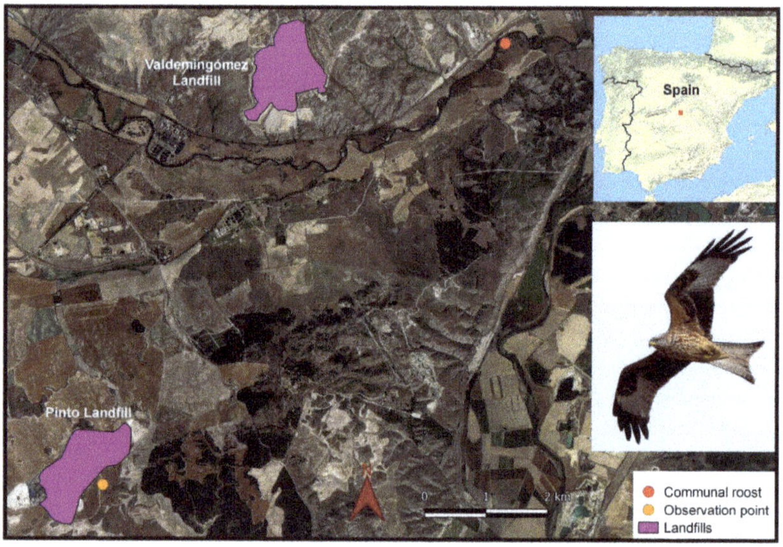

Figure 1. Map showing the location of landfills, observation point, and communal roost of the red kite in southeastern Madrid, central Spain.

Individuals of other bird species present in the landfill were counted using the same methodology. Only medium and large species were considered due to the impossibility of detecting all individuals of species smaller than the spotless starling (*Sturnus unicolor*); the number of individuals of the latter species could be estimated due to the continuous flights over the rubbish.

To assess the flow rate of red kites entering and leaving the landfill, counts of individuals crossing the perimeter of the landfill in the direction of entry and exit were made for 15 min. These surveys were conducted every hour and took place just after each count of the total number of individuals present at the landfill. The data obtained by sampling the flow of individuals were transformed in terms of absolute flow (or turnover rate), calculated as the difference between the number of individuals entering and leaving the landfill in each sampling period.

2.4. Age Determination

The age (juvenile or adult) of red kites foraging at the landfill was assessed by photographs of randomly observed individuals across the study period ($n = 218$), taken with a digital camera (Sony alpha 6600, with FE 200–600 mm lens). The age of a sample of individuals ($n = 179$), observed with a telescope in the trees that serve as communal roosts in the area, was determined as an approximation to the age distribution in the population as a whole; two snapshot samplings were conducted in the communal roost on 27 January and 2 February 2021. Individuals were classified as juveniles (born in the previous breeding season) or adults (born in previous years) by plumage characteristics and iris color [60].

2.5. Diet

The diet of the wintering population was evaluated through the analysis of pellets ($n = 71$) collected in January 2021 in a communal roost located on the riverine forest at the banks of the Manzanares River, about 4 km from the Valdemingómez and 10 km from the Pinto landfills, respectively (Figure 1). Prey remains contained in the pellets were macroscopically identified from reference collections. Food remains were classified according to their origin in the landfills versus those obtained in the countryside. The food obtained from landfills included any domestic livestock remains, such as poultry, pig, cow, lamb, and marine fish, as this type of food can only be obtained from household waste (hereafter, "organic waste") due to the absence of specific dumps of livestock carcasses in the study area. The presence of anthropogenic debris such as plastics, paper, metal, and glass was also recorded in the pellets, reflecting the use of landfills for foraging, and was categorized as "synthetic litter". The remains from wild animals were classified into four categories, including wild rabbits (*Oryctolagus cuniculus*), other wild mammals, birds, and insects.

To analyze the diet, the percentage of pellets in which each food category appeared was used, as in other studies on the diet of this and other scavenger species [22,61]. The pellets of this species generally consist of the hair or feathers of the consumed animals. This makes it impossible to quantify the minimum number of specimens consumed due to the difficult-to-control variations derived from the consumption of individual animal fragments obtained by scavenging, which also prevents the quantification of the biomass contributed by each type of food, as occurs in other scavenger species [62].

2.6. Data Analysis

We used Generalized Linear Mixed Models (GLMMs) to investigate factors affecting the number of red kites using the landfill in each snapshot count (Poisson error distribution, log-link function) and the flow of individuals entering and leaving the landfill (turnover rate) in each sampling period (Gaussian error distribution, identity link function). Explanatory variables included daylight period (morning, midday, and afternoon), month, and the interaction of both factors. In the analysis of the turnover rate, the number of individuals present in the landfill in the count just before the estimation of the flow of individuals was also included as a covariate. We included the day of the counts as a random term in both models, as the number of kites and the turnover rate could be similar within days due to specific environmental and social conditions. The age distribution (juveniles/adults) of individuals foraging in the landfill was compared with that of the overall population (sampled at communal roosts) using Fisher's exact test. Statistical analyses were performed using SPSS software v. 28 (IBM SPSS Statistics, Armonk, NY, USA). Statistical significance was set at $p < 0.05$ (2-sided).

3. Results

3.1. Abundance of Red Kites and Other Species Foraging in the Landfill

We counted individuals of 16 medium–large bird species in the landfill (Table 1). The red kite was observed feeding at the landfill on all sampling days and was present in the vast majority of daily counts. Other frequent species were the black-backed gull (*Larus fuscus*), black-headed gull (*Chroicocephalus ridibundus*), and the white stork (*Ciconia ciconia*), which also had the highest mean abundances and maximum counts. Comparatively, the red kite showed a low mean abundance, with a maximum of 92 individuals recorded in one of the hourly counts. Other species were observed every day but in a lower proportion of the hourly counts, while the remaining species can be considered infrequent visitors in low abundance, as with the presence of endangered species such as the cinereous vulture (*Aegypius monachus*) or rarities such as the great egret (*Ardea alba*). It is worth noting the very scarce presence of some wintering black kites (Table 1). Several smaller species were recorded feeding at the landfill but could not be adequately counted, including the house sparrow (*Passer domesticus*), the white wagtail (*Motacilla alba*), and the domestic pigeon (*Columba livia*).

3.2. Seasonal and Hourly Abundance and Flow of Red Kites

According to the GLMM, the number of red kites recorded in the landfill was influenced by the daylight period and its interaction with the month (Table 2). This indicates that the number of red kites was lower in the morning, while the number of kites as the day progressed depended on the month, being highest at midday in December and similar between midday and afternoon in January and February (Figure 2).

Table 1. Bird species foraging at the Pinto landfill, southeastern Madrid, central Spain, during the winter of 2021. Data show the frequency of occurrence on sampling days and counts, their mean and maximum abundances, and the sum of all counts.

Species	% Days $n = 11$	% of Counts $n = 189$	Mean ± SD Abundance	Maximum	Sum
Milvus milvus	100	99.5	27 ± 19	92	5164
Larus fuscus	100	94.2	1531 ± 1749	9274	289,330
Chroicocephalus ridibundus	100	91.5	37 ± 43	226	7048
Ciconia ciconia	100	89.9	65 ± 87	620	12,270
Corvus corax	100	32.3	0.8 ± 1.6	9	154
Sturnus unicolor	100	31.7	53 ± 120	800	10,036
Buteo buteo	100	26.5	0.3 ± 0.6	3	59
Falco tinnunculus	81.8	19.6	0.2 ± 0.5	3	45
Pica pica	45.5	10.1	0.4 ± 1.3	7	78
Circus aeruginosus	63.6	6.3	0.06 ± 0.2	1	12
Aegypius monachus	54.5	5.8	0.07 ± 0.3	2	14
Bubulcus ibis	36.4	4.8	0.2 ± 1.3	14	38
Milvus migrans	45.5	4.2	0.04 ± 0.2	1	8
Corvus monedula	27.3	4.2	0.5 ± 3.7	40	97
Accipiter nisus	36.4	2.1	0.2 ± 0.1	1	4
Ardea alba	9.1	0.5	0.005 ± 0.07	1	1

Table 2. Results of the GLMM conducted to evaluate factors affecting the number of red kites in each snapshot count in the Pinto landfill, southeastern Madrid, central Spain. Explanatory variables included daylight period (morning, midday, afternoon), month, and the interaction of both factors.

Predictors	Estimate	SE	*t*	*p*	CI (95%) Inf.	CI (95%) Sup.
(intercept)	3.614	0.4393	8.227	<0.001	2.747	4.481
Period = morning	−0.409	0.1006	−4.063	<0.001	−0.607	−0.210
Period = midday	0.119	0.0857	1.393	0.165	−0.050	0.289
Month = December	−0.430	0.4702	−0.914	0.362	−1.358	0.498
Month = January	−0.590	0.5087	−1.160	0.247	−1.594	0.413
[Period = morning] × [month = December]	0.538	0.1097	4.906	<0.001	0.322	0.754
[Period = morning] × [month = January]	−0.040	0.1332	−0.302	0.763	−0.303	0.222
[Period = midday] × [month = December]	0.118	0.0947	1.243	0.216	−0.069	0.305
[Period = midday] × [month = January]	−0.126	0.1123	−1.118	0.265	−0.347	0.096

Analysis of the turnover rate shows no significant influence of the daylight period, month, number of individuals present in the landfill in the count just before the estimation of the turnover rate, and the interaction between these variables (GLMM, all $p > 0.17$). Pooling all data, the mean numbers of individuals entering and leaving the landfill in each sampling period ($n = 96$) were very similar (9.4, SD = 7.4 and 9.2, SD = 7.2, respectively).

3.3. *Age Distribution*

The proportion of kites of each age class recorded in the communal roosts was not statistically different between the sampling in January (79.1% of adults, $n = 91$) and February (84.1% of adults, $n = 88$) (Fisher's exact test, $p = 0.4437$). Overall, the proportion of each age class was different between the landfill (63.3% adults, 36.7% juveniles, $n = 218$) and the total population sampled in the communal roost (81.5% adults, 18.5% juveniles, $n = 179$, pooling both samplings) (Fisher's exact test, $p < 0.001$).

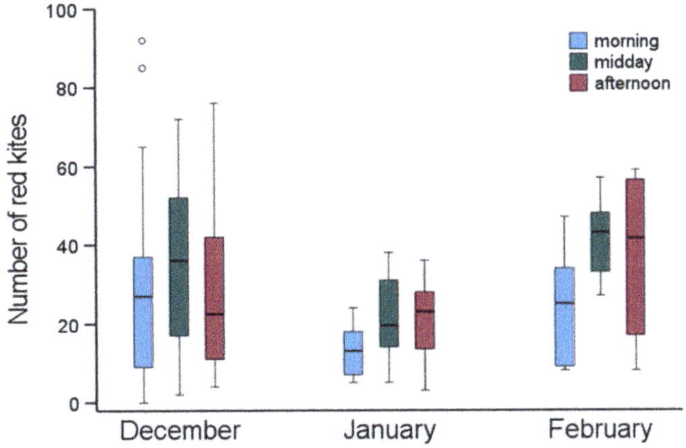

Figure 2. Median, interquartile range, outliers, and extreme cases of the number of red kites recorded in the Pinto landfill in southeastern Madrid, central Spain, according to daylight period (morning, midday, and afternoon) in each study month in the winter of 2020–2021.

3.4. Diet

The quantification of the food remains found in pellets ($n = 71$) is shown in Figure 3. Most pellets (97.2%) contained remains of wild rabbits. Remains of other wild animals were present in a very low proportion of pellets and correspond to mammals, including wild boar (*Sus scrofa*, 1.4%), house mouse (*Mus musculus*, 2.8%), Mediterranean pine vole (*Microtus duodecimcostatus*, 1.4%), birds (domestic pigeon, 1.4%), and insects (unidentified beetles and ants; 5.6%). The occurrences classified as "carrion obtained from landfill" corresponded to bones and feathers of poultry (*Gallus gallus domesticus*, 4.2%), while pellets with synthetic rubbish included plastic fragments and a small piece of metal from a commercially available sausage remnant (1.4%).

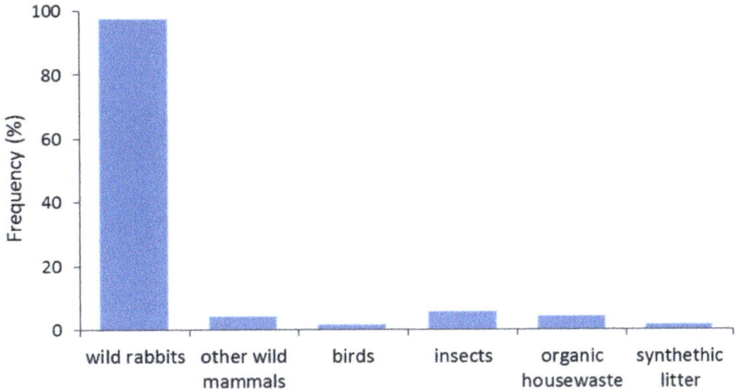

Figure 3. Proportion of occurrence of each food type in pellets ($n = 71$) of red kite, collected in the communal roost near the landfills of southeastern Madrid, central Spain, during the study period (winter 2020–2021).

Because each pellet may contain remains of different food types, the sum of the proportions of each of the considered food categories can add up to more than 100%. Thus, most of the pellets were composed exclusively of wild rabbit hair and bones (61 of 71, 85.9%), while a large majority of the pellets with other food types also contained rabbit remains (8 of 10, 80.0%). Only two of the analyzed pellets (2.8%) contained no rabbit remains: one consisted exclusively of poultry feathers and the other of poultry feathers and insect remains. The only pellet with synthetic litter consisted mostly of rabbit remains.

4. Discussion

Around 17% of the world's food production is wasted annually, which translates into a global average of 121 kg of household waste per capita [63]. The dumping of household waste in open landfills is still a common practice in many European regions, and it is expected to last in forthcoming decades [3,55]. In Spain, around 17.5 million tons of food wasted annually are primarily disposed of in open landfills, where red kites forage on a daily or seasonal basis, especially during winter, thus exerting a generally unknown influence on their populations. Our results show the constant presence of red kites throughout the day and during the whole winter period at a landfill in central Spain. Gulls of two species and the white stork also frequently used the landfill and showed much higher abundances than red kites, while several other scavengers and generalist foragers were frequently observed but at much lower abundances. The presence during the winter of red kites in landfills located in southeastern Madrid was not recorded in the past [17,64,65], and the use of the area as wintering grounds, including the establishment of communal roosts, is a recent event.

Nowadays, seasonal migratory movements dictate the variations in the number of red kites foraging in the landfill during the winter, with a slightly higher abundance in December and February than in January. This suggests that individuals moving to more southerly latitudes can spatiotemporally overlap with local and migratory individuals spending the winter in the study area [20,66]. In addition, individuals returning to their breeding areas from southern latitudes can coincide with those that have not yet begun their migration towards the north, thus increasing their abundance in February in central Spain [34,44,67]. These seasonal movements, together with the typical nomadic movements during the wintering season, suggest that the recent use of landfills could play a role in the movements of wandering and migrating red kites, as well as for those established in the area as their wintering and breeding grounds.

The number of red kites in the landfill increased until midday and then decreased as individuals returned to the roost, although with variations depending on the month. This agrees with similarly frequent entries in the morning and in the afternoon, and some accumulation of individuals during the central part of the day. The lack of any clear daylight turnover pattern in the landfill may be influenced by kites' activities that are not related to feeding but to social interactions, including chases, fights, food thefts, and passive gatherings that can be envisioned within the range of interactions with a social function in this species [21,36]. Specifically, it was frequently observed that after obtaining food remains, individual kites quickly left the landfill followed by other individuals attempting to steal the food. The landfill was also used during midday as a resting place after feeding and as a place of concentration before returning to the roost. Therefore, the counts of red kites could overestimate the landfill's importance as foraging grounds, as not all individuals observed necessarily feed there. In addition, kites also predate on wild animals in the landfill (small birds and rodents) and their close surroundings, especially wild rabbits breeding and foraging at a high density in the embankments on the edges of the landfill and surrounding crops and hillsides.

Generalist predators and scavengers may select particular food and foraging places depending on energetic and social constraints [68,69]. Because red kites are opportunistic predators and facultative scavengers, their diet includes a high variety of items, exploited based on abundance, availability, predictability, and preference [21,22,33,35,70]. The results of the pellet analysis make it clear that the diet of red kites in the study area is dominated by the wild rabbit, a key prey species in the Mediterranean [71]. This dependence is favored by the high abundance of this prey in the study area and their surroundings, now and in the past [22,33], which indicates that the natural environment offers a much greater quantity of suitable food than the landfill itself. Food scraps obtained from a landfill may leave little residue in the pellets, which would lead to underestimating their importance in the diet [33]. While this may be true for meat scraps without integument, it does not apply to household waste and slaughterhouse scraps and items that may result in the presence of small bones, feathers, or scales, as in the case of poultry and fish scraps. In addition, the remains obtained from the landfill are associated with indigestible synthetic debris, such as plastic, paper, small glass, and metal fragments, which should be present in a higher proportion of the pellets if the landfills were an important source of food at the population level. Even then, this would not invalidate the preponderant importance of the wild rabbit in the diet. In any case, the potential underestimation of the importance of the remains obtained from the landfill would have to be associated with the simultaneous consumption of rabbits, suggesting that the role of this prey item is quantitatively very relevant compared to the food obtained from the landfill, as occurred in the black kite in the same area [37].

The wild rabbit represents the ideal prey type for many medium- and large-sized raptors and carnivore mammals in the Mediterranean, especially when this key prey species reached a high density [71]. Other avian species with a much higher abundance in the landfill, such as gulls and the white stork, can compete with red kites for food scraps [15,36], whereas when feeding on wild prey, this competition does not occur. Therefore, the importance of landfills may be greater as predictable feeding sites than as places where food may be available as an abundant, easily accessible, defensible, and nutritionally high-quality resource. In particular, the scraps generally obtained from household waste are small, low energy, and low nutritional value items, compared with the wild prey and carrion exploited in livestock carcass dumps [15,22,25,37,40]. However, due to their predictability, food scraps obtained in landfills can be important in periods of scarce availability of wild prey or bad weather. These resources could be especially valuable for young and inexperienced, migrating or wandering individuals with little knowledge of the study area, and experience and ability to search for and capture more optimal food-animal, both as live prey or carcass, as well as for uncompetitive individuals or those in poor physical condition. This is in agreement with a higher proportion of juveniles in the landfill than in the overall wintering population, estimated by sampling at roosts. In addition, the landfill could act as a place with a social function that could be extended to nearby communal roosts [72].

The wintering population in the study area was established in 2009 and has increased progressively since then, reaching about 3000 individuals estimated through simultaneous counts in communal roosts in the last years (authors' unpubl. data), which represents one of the numerically most important wintering nuclei in Spain. This contrasts with the relatively low abundance of kites observed in the studied landfill, with less than 30 individuals on average across daily snapshot counts throughout the winter and a reduced hourly inflow and outflow. The use of the other landfill in the area (Valdemingómez) could be assumed similar, although the waste management by recycling and incineration there [55] can make food resources less available for scavengers. Moreover, in the last few years, campaigns have been launched for reasons unknown to us to scare away the birds that feed in this landfill (pers. obs.) by those responsible for waste management by the regional governments.

5. Conclusions

Our study suggests relatively low quantitative importance of landfills as foraging grounds on a daily basis for the increasing population of wintering red kites in southeastern Madrid. Instead, the high regional density of wild rabbits, and in the surroundings of the landfills, attracts large numbers of red kites that can eventually use these places, thus likely influencing an uninformed positive perception about its relevance for the conservation of the wintering population. In contrast to predation on wild rabbits, the use of the landfill as a foraging site and of the refuse as food is subject to multiple risks to the health and survival of the kites. These risks include mainly poisoning and contamination through ingestion of toxic products and hazardous synthetic materials, electrocution and collision with wires and fences, entanglement with ropes and plastics, and unintentional poisoning with anticoagulants used in rat extermination campaigns. More research is needed to evaluate the role of increasing populations of wild rabbits on the habitat use and population dynamics of the red kite in central Spain.

Author Contributions: Conceptualization, G.B; methodology, Í.V.-H., F.M. and G.B.; formal analysis, Í.V.-H. and G.B.; investigation, Í.V.-H., F.M. and G.B.; resources, Í.V.-H., F.M. and G.B.; writing—original draft preparation, Í.V.-H. and G.B.; writing—review and editing, Í.V.-H., F.M. and G.B.; supervision, G.B.; project administration, G.B.; funding acquisition, Í.V.-H., F.M. and G.B.; All authors have read and agreed to the published version of the manuscript.

Funding: Fieldwork was funded by the project PID2019-109685GB-I00 of the Spanish Ministry of Science and Innovation.

Institutional Review Board Statement: Not applicable.

Data Availability Statement: The datasets are available from the corresponding author on reasonable request.

Conflicts of Interest: The authors declare no conflict of interest.

References

1. Gordon, A.T. *Solid Waste Management*; MC Graw Hill: New York, NY, USA, 2000.
2. Schanes, K.; Dobernig, K.; Gözet, B. Food waste matters—A systematic review of household food waste practices and their policy implications. *J. Clean. Prod.* **2018**, *182*, 978–991. [CrossRef]
3. Chen, D.M.-C.; Bodirsky, B.L.; Krueger, T.; Mishra, A.; Popp, A. The world's growing municipal solid waste: Trends and impacts. *Environ. Res. Lett.* **2020**, *15*, 074021. [CrossRef]
4. Brems, A.; Baeyens, J.; Dewil, R. Recycling and recovery of post-consumer plastic solid waste in a European context. *Therm. Sci.* **2012**, *16*, 669–685. [CrossRef]
5. zu Ermgassen, E.K.H.J.; Balmford, A.; Salemdeeb, R. Reduce, relegalize, and recycle food waste. *Science* **2016**, *352*, 1526. [CrossRef] [PubMed]
6. Bugge, M.M.; Fevolden, A.M.; Klitkou, A. Governance for system optimization and system change: The case of urban waste. *Res. Policy* **2019**, *48*, 1076–1090. [CrossRef]
7. Kutsevych, M.; Yara, O.; Golovko, L.; Terpeliuk, V. Sustainable Approaches to Waste Management: Regulatory and Financial Instruments. *Eur. J. Sustain. Dev.* **2020**, *9*, 163–171. [CrossRef]
8. Magrini, C.; Degli Esposti, A.; De Marco, E.; Bonoli, A. A framework for sustainability assessment and prioritisation of urban waste prevention measures. *Sci. Total Environ.* **2021**, *776*, 145773. [CrossRef]
9. Romano, G.; Marciano, C.; Fiorelli, M.S. *Urban Waste Management in Europe: Challenges and Opportunities, Best Practices in Urban Solid Waste Management*; Emerald Publishing Limited: Bingley, UK, 2021; pp. 9–44. [CrossRef]
10. Slack, R.; Gronow, J.; Voulvoulis, N. Hazardous components of household waste. *Crit. Rev. Environ. Sci. Technol.* **2004**, *34*, 419–445. [CrossRef]
11. Giusti, L. A review of waste management practices and their impact on human health. *Waste Manag.* **2009**, *29*, 2227–2239. [CrossRef]
12. Blanco, G.; Montoya, R. Landscape, "muladares", or poisoning? Using GIS to assess factors related to the decline of breeding red kites (*Milvus milvus*) in Spain. In *Spatial Analysis in Raptor Ecology and Conservation*; Rodríguez Estrella, R.B.T., Ed.; CIBNOR & Comisión Nacional para el Conocimiento y Uso de la Biodiversidad: México City, México, 2004; pp. 133–152.
13. Plaza, P.I.; Lambertucci, S.A. How are garbage dumps impacting vertebrate demography, health, and conservation? *Glob. Ecol. Conserv.* **2017**, *12*, 9–20. [CrossRef]
14. Noreen, Z.; Sultan, K. A global modification in avifaunal behavior by use of waste disposal sites (waste dumps/rubbish dumps): A review paper. *Pure Appl. Biol.* **2021**, *10*, 603–616. [CrossRef]
15. Donázar, J.A. Muladares y basureros en la biologia y conservacion de las aves en España. *Ardeola* **1992**, *39*, 29–40.
16. BirdLife International. Species Factsheet: *Milvus milvus*. 2021. Available online: http://www.birdlife.org (accessed on 7 August 2021).
17. Viñuela, J.; Martí, R.; Ruiz, A. *El Milano Real en España. Monografía no.6*; SEO/BirdLife: Madrid, Spain, 1999.
18. Literák, I.; Horal, D.; Alivizatos, H.; Matušík, H. Common wintering of black kites (*Milvus migrans migrans*) in Greece, and new data on their wintering elsewhere in Europe. *Slovak Raptor J.* **2017**, *11*, 91–102. [CrossRef]
19. Literák, I.; Horal, D.; Raab, R.; Matušík, H.; Vyhnal, S.; Rymešová, D.; Spakovszky, P.; Skartsi, T.; Poirzidis, K.; Zakkak, S.; et al. Sympatric wintering of red kites and black kites in southeast Europe. *Acta Zool. Acad. Sci. Hung.* **2019**, *65*, 381–398. [CrossRef]
20. Panter, C.T.; Literák, I.; Raab, R.; Tolhurst, B.A.; White, R.L. Age, landscape, and arrival date explain ranging behavior of wintering red kites in southwest Europe. *J. Wildl. Manag.* **2022**, *86*, e22147. [CrossRef]

21. Mougeot, F.; Garcia, J.T.; Viñuela, J. Breeding biology, behaviour, diet and conservation of the red kite (*Milvus milvus*), with particular emphasis on Mediterranean populations. In *Ecology and Conservation of European Dwelling Forest Raptors and Owls*; Zuberogitia, I., Martínez, J.E., Eds.; Editorial Diputación Foral de Vizcaya: Bilbao, Spain, 2011; pp. 190–204.
22. Blanco, G. Can livestock carrion availability influence diet of wintering red kites? Implications of sanitary policies in ecosystem services and conservation. *Popul. Ecol.* **2014**, *56*, 593–604. [CrossRef]
23. Blanco, G. Influence of diet on the gastrointestinal flora of wintering red kites. *Eur. J. Wildl. Res.* **2014**, *60*, 695–698. [CrossRef]
24. Cortés-Avizanda, A.; Blanco, G.; De Vault, T.L.; Markandya, A.; Virani, M.Z.; Donázar, J.A. Supplementary feeding and endangered avian scavengers: Benefits, caveats and controversies. *Front. Ecol. Environ.* **2016**, *14*, 191–199. [CrossRef]
25. Blanco, G.; Cardells, J.; Garijo-Toledo, M.M. Supplementary feeding and endoparasites in threatened avian scavengers: Coprologic evidence from red kites in their wintering stronghold. *Environ. Res.* **2017**, *155*, 22–30. [CrossRef]
26. Blanco, G.; de Tuesta, J.A.D. Culture- and molecular-based detection of swine-adapted Salmonella shed by avian scavengers. *Sci. Total Environ.* **2018**, *634*, 1513–1518. [CrossRef]
27. Viada, C. La milana reial (*Milvus milvus*) a Mallorca. *Bolletí De La Soc. D'història Nat. De Les Balear.* **1994**, *37*, 101–107.
28. Contreras, A. *Impacto Sobre La Avifauna De La Implantación Del Plan De Gestión De Los Residuos Sólidos Urbanos En Segovia*; Caja Segovia: Segovia, Spain, 2001.
29. Camiña, Á.; Barrio, E.M. Evolución estacional de las aves no passeriformes asociadas al vertedero de RSU de Nájera. *Zubía* **2005**, *23*, 7–22.
30. Evans, J.M.; Pimkowski, M.W. World status of the Red Kite. *Br. Birds* **1991**, *1*, 171–187.
31. Knott, J.; Newbury, P.; Barov, B. *Action Plan for the Red Kite Milvus milvus in the European Union*; BirdLife International for the European Union: Cambridge, UK, 2009.
32. Molina, B. *El Milano Real En España. III Censo Nacional. Población Invernante y Reproductora En 2014 y Métodos De Censo*; SEO/Birdlife: Madrid, Spain, 2015.
33. Ortega, A.; Casado, S. Alimentación del Milano Real Milvus milvus en la provincia de Madrid. *Doñana Acta Vertebr.* **1991**, *18*, 195–204.
34. Sunyer, C.; Viñuela, J. Variación temporal en los hábitos alimentarios del Milano Real durante la invernada en la Meseta Norte. *Ardeola* **1994**, *41*, 161–167.
35. García, J.T.; Viñuela, J.; Sunyer, C. Geographic variation of the winter diet of the Red Kite *Milvus milvus* in the Iberian Peninsula. *IBIS* **1998**, *140*, 302–309. [CrossRef]
36. Gómez-Tejedor, U. Comportamiento Cleptopárasitico del Milano Real *Milvus milvus* en un Vertedero. In *Holarctic Birds of Prey: Proceedings of an International Conference*; Chancellor, R.D., Meyburg, B.-U., Ferrero, J.J., Eds.; Actas del Congreso International sobre Rapaces del Holártico: Mèrida, Spain; Berlin, Germany, 1998; pp. 173–176.
37. Blanco, G. Role of refuse as food for migrant, floater and breeding Black Kites (*Milvus migrans*). *J. Raptor Res.* **1997**, *31*, 71–76.
38. De Giacomo, U.; Guerrieri, G. The Feeding Behavior of the Black Kite (*Milvus migrans*) in the Rubbish Dump of Rome. *J. Raptor Res.* **2008**, *42*, 110–118. [CrossRef]
39. Mazumdar, S.; Ghose, D.; Saha, G.K. Preferences for Different Types of Offal by Black Kites *Milvus migrans* from Urban Garbage Dumps of Kolkata, India. *Acta Ornithol.* **2019**, *53*, 163. [CrossRef]
40. Serrano, D. Dumps for dead livestock and the conservation of wintering red kites (*Milvus milvus*). *J. Raptor Res.* **1999**, *33*, 338–340.
41. BirdLife International. *Milvus milvus*. The IUCN Red List of Threatened Species 2020: E.T22695072A181651010. Available online: https://www.iucnredlist.org/species/22695072/181651010 (accessed on 6 August 2021).
42. Orros, M.E.; Fellowes, M.D.E. Widespread supplementary feeding in domestic gardens explains the return of reintroduced Red Kites *Milvus milvusto* an urban area. *IBIS* **2015**, *157*, 230–238. [CrossRef] [PubMed]
43. Cereghetti, E.; Scherler, P.; Fattebert, J.; Grüebler, M.U. Quantification of anthropogenic food subsidies to an avian facultative scavenger in urban and rural habitats. *Landsc. Urban Plan.* **2019**, *190*, 103606. [CrossRef]
44. Blanco, G.; Frías, O.; Arroyo, B.; Martínez, F.; Baniandrés, N. Contamination traps as trans-frontier management challenges: New research on the impact of refuse dumps on the conservation of migratory avian scavengers. In *Environmental Research Trends*; Nova Science Publishers: New York, NY, USA, 2007; pp. 153–204.
45. Smart, J.; Amar, A.; Sim, I.M.; Etheridge, B.; Cameron, D.; Christie, G.; Wilson, J.D. Illegal killing slows population recovery of a re-introduced raptor of high conservation concern—The red kite *Milvus milvus*. *Biol. Conserv.* **2010**, *143*, 1278–1286. [CrossRef]
46. Tenan, S.; Adrover, J.; Muñoz-Navarro, A.; Sergio, F.; Tavecchia, G. Demographic consequences of poison-related mortality in a threatened bird of prey. *PLoS ONE* **2012**, *7*, e49187. [CrossRef]
47. Mateo-Tomás, P.; Olea, P.P.; Mínguez, E.; Mateo, R.; Viñuela, J. Direct evidence of poison-driven widespread population decline in a wild vertebrate. *Proc. Natl. Acad. Sci. USA* **2020**, *117*, 16418–16423. [CrossRef]
48. Schaub, M. Spatial distribution of wind turbines is crucial for the survival of red kite populations. *Biol. Conserv.* **2012**, *155*, 111–118. [CrossRef]
49. Bellebaum, J.; Korner-Nievergelt, F.; Dürr, T.; Mammen, U. Wind turbine fatalities approach a level of concern in a raptor population. *J. Nat. Conserv.* **2013**, *21*, 394–400. [CrossRef]
50. Crespo-Luengo, G.; Hernández-Lambraño, R.E.; Barbero-Bermejo, I.; Sánchez-Agudo, J. Analysis of Spatio-Temporal Patterns of Red Kite *Milvus milvus* Electrocution. *Ardeola* **2020**, *67*, 247–268. [CrossRef]

51. Serrano, D.; Margalida, A.; Pérez-García, J.M.; Juste, J.; Traba, J.; Valera, F.; Carrete, M.; Aihartza, J.; Real, J.; Mañosa, S.; et al. Renewables in Spain threaten biodiversity. *Science* **2020**, *370*, 1282–1283. [CrossRef]
52. Blanco, G.; Sergio, F.; Frías, Ó.; Salinas, P.; Tanferna, A.; Hiraldo, F.; Barceló, D.; Eljarrat, E. Integrating population connectivity into pollution assessment: Overwintering mixing reveals flame retardant contamination in breeding areas in a migratory raptor. *Environ. Res.* **2018**, *166*, 553–561. [CrossRef]
53. Blanco, G.; Gómez-Ramírez, P.; Espín, S.; Sánchez-Virosta, P.; Frías, Ó.; García-Fernández, A.J. Domestic Waste and Wastewaters as Potential Sources of Pharmaceuticals in Nestling White Storks (*Ciconia ciconia*). *Antibiotics* **2023**, *12*, 520. [CrossRef]
54. Comunidad de Madrid. *Estrategia de Residuos de la Comunidad de Madrid (2006–2016)*; Consejería de Medio Ambiente y Ordenación del Territorio: Madrid, Spain, 2007.
55. Comunidad de Madrid. *Estrategia de Gestión Sostenible de los Residuos de la Comunidad de Madrid (2017–2024)*; Consejería de Medio Ambiente y Ordenación del Territorio: Madrid, Spain, 2018.
56. *Comunidad de Madrid El Parque Regional del Sureste Madrileño*; Consejería de Medio Ambiente y Desarrollo Regional: Madrid, Spain, 1999.
57. Blanco, G.; Jiménez, B.; Frías, O.; Millan, J.; Dávila, J.A. Contamination with nonessential metals from a solid-waste incinerator correlates with nutritional and immunological stress in prefledgling black kites (*Milvus migrans*). *Environ. Res.* **2004**, *94*, 94–101. [CrossRef] [PubMed]
58. De Sanctis, A.; Mariottini, M.; Fanello, E.L.; Blanco, G.; Focardi, S.E.; Guerranti, C.; Perra, G. Evaluating contamination in the red-billed chough *Pyrrhocorax pyrrhocorax* through non-invasive sampling. *Microchem. J.* **2013**, *107*, 70–75. [CrossRef]
59. Pitarch, A.; Diéguez-Uribeondo, J.; Martín-Torrijos, L.; Sergio, F.; Blanco, G. Fungal signatures of oral disease reflect environmental degradation in a facultative avian scavenger. *Sci. Total Environ.* **2022**, *837*, 155397. [CrossRef] [PubMed]
60. Blasco-Zumeta, J.; Heinze, G.-M. *Atlas de Identificación de las Aves Continentales de la Península Ibérica*; Tundra Ediciones: Almenara, Spain, 2022.
61. Donázar, J.A.; Cortés-Avizanda, A.; Carrete, M. Dietary shifts in two vultures after the demise of supplementary feeding stations: Consequences of the EU sanitary legislation. *Eur. J. Wildl. Res.* **2010**, *56*, 613–621. [CrossRef]
62. Blanco, G.; Hornero-Méndez, D. Interspecific differences in plasma carotenoid profiles in nestlings of three sympatric vulture species. *Curr. Zool.* **2023**, zoac090. [CrossRef]
63. UNEP. *Food Waste Index Report 2021*; United Nations Environment Programme: Nairobi, Kenya, 2021.
64. Blanco, G. Population Dynamics and Communal Roosting of White Storks Foraging at a Spanish Refuse Dump. *Colon. Waterbirds* **1996**, *19*, 273–276. [CrossRef]
65. Cardiel, I.E. *El Milano Real (Milvus milvus) en España. II Censo Nacional*; SEO/BirdLife: Madrid, Spain, 2005.
66. Literák, I.; Raab, R.; Škrábal, J.; Vyhnal, S.; Dostál, M.; Matušík, H.; Makoň, K.; Maderič, B.; Spakovszky, P. Dispersal and Philopatry in Central European Red Kites *Milvus milvus*. *J. Ornithol.* **2022**, *163*, 469–479. [CrossRef]
67. Pfeiffer, T.; Meyburg, B.U. Migratory and wintering behaviour of the Red Kite *Milvus milvus* in Thuringia (Germany) as revealed by satellite telemetry. *Vogelwarte* **2009**, *47*, 171–187.
68. Street, G.M.; Fieberg, J.; Rodgers, A.R.; Carstensen, M.; Moen, R.; Moore, S.A.; Windels, S.K.; Forester, J.D. Habitat functional response mitigates reduced foraging opportunity: Implications for animal fitness and space use. *Landsc. Ecol.* **2016**, *31*, 1939–1953. [CrossRef]
69. van Overveld, T.; Blanco, G.; Moleón, M.; Margalida, A.; Sánchez-Zapata, J.A.; de la Riva, M.; Donázar, J.A. Integrating vulture social behavior into conservation practice. *Condor* **2020**, *122*, duaa035. [CrossRef]
70. Sergio, F.; Chicano, J.; Tanferna, A.; Blas, J.; Hiraldo, F. El papel de los Parques Nacionales en la protección de especies amenazadas: Declive y demografía del Milano Real en el Espacio Natural de Doñana y en Andalucía a lo largo de cuatro décadas. *Proy. De Investig. En Parq. Nac.* **2021**, *2015–2019*, 301–313.
71. Delibes-Mateos, M.L.; Delibes, M.; Ferreras, P.; Villafuerte, R. Key role of European rabbits in the conservation of the Western Mediterranean basin hotspot. *Conserv. Biol.* **2008**, *22*, 1106–1117. [CrossRef]
72. Hiraldo, F.; Heredia, B.; Alonso, J.C. Communal Roosting of Wintering Red Kites *Milvus milvus* (Aves, Accipitridae): Social Feeding Strategies for the Exploitation of Food Resources. *Ethology* **1993**, *93*, 117–124. [CrossRef]

diversity

MDPI

Article

Turnover and Natal Dispersal in the Finnish Golden Eagle (*Aquila chrysaetos*) Population

Aure Kylmänen [1], Ekaterina Karabanina [1,*], Tuomo Ollila [2], Suvi Ponnikas [1] and Laura Kvist [1]

1 Ecology and Genetics Research Unit, University of Oulu, 90014 Oulu, Finland; aurekylmanen@gmail.com (A.K.); suvi.ponnikas@oulu.fi (S.P.); laura.kvist@oulu.fi (L.K.)
2 Metsähallitus, Natural Heritage Services, 96101 Rovaniemi, Finland; tuomo.ollila@metsa.fi
* Correspondence: ekaterina.karabanina@oulu.fi

Abstract: Estimating turnover in a population provides information on population dynamics, such as dispersal and mortality. Dispersal increases genetic diversity and affects the genetic structure. Golden eagles are monogamous, tend to mate for life, and have strong nest site fidelity, which suggests low turnover rates. Here, we first studied genetic diversity and population structure in the Finnish golden eagle population using 11 microsatellite loci and a fragment of a mitochondrial DNA control region. We found no notable changes in genetic diversity during the 15-year study period and did not discover any population structure. Then, we examined the turnover rate using chick genotypes (N = 935) by estimating relatedness between chicks born in the same territory in different years. The results showed a turnover rate of 23%, which correlated with the breeding success of the previous year. Similarly, in the absence of turnover, the pair changed nest sites within a territory after an unsuccessful breeding. In addition, our dataset also revealed natal dispersal of ten individuals. Natal dispersal distance was 110 km on average (median 98 km); however, the distance seemed to vary depending on geographical location, being greater in Northern Finland than in Southern Finland.

Keywords: raptor ecology; breeding dispersal; microsatellites; mtDNA

Citation: Kylmänen, A.; Karabanina, E.; Ollila, T.; Ponnikas, S.; Kvist, L. Turnover and Natal Dispersal in the Finnish Golden Eagle (*Aquila chrysaetos*) Population. *Diversity* **2023**, *15*, 567. https://doi.org/10.3390/d15040567

Academic Editors: Vladimir D. Dobrev, Dobromir D. Dobrev, Dimitar A. Demerdzhiev and Michael Wink

Received: 31 March 2023
Revised: 12 April 2023
Accepted: 14 April 2023
Published: 18 April 2023

1. Introduction

Dispersal has significant implications for genetic diversity and structure of populations, as it acts as a mechanism for gene flow [1]. Dispersal increases genetic diversity and decreases inbreeding in populations, which makes it imperative for species of conservation concern. Dispersal also allows for colonization of new habitats and expansion of the current range [1]. Despite its importance, dispersal can be challenging to study, especially in long-lived and highly mobile species, which prefer to live remotely from humans [2]. In addition, it is difficult to distinguish between mortality and dispersal, because the individuals may not be found once dead, or will not be observed once they have left the study area [3].

Breeding dispersal refers to movement of individuals between breeding sites, whereas natal dispersal refers to movement between the place of birth and the first breeding site [4]. Natal dispersal tends to be more extensive than breeding dispersal, possibly because juveniles leave natal nest sites to avoid breeding or competing with relatives [4,5]. Thus, the two primary hypotheses that describe the potential factors affecting natal dispersal are intraspecific competition over resources and inbreeding avoidance [1]. Especially, species with high survival rates and philopatry to breeding sites might disperse further to find areas with lower population density and, therefore, less competition [1]. However, if the natal site has high primary productivity, the dispersal distance may be shorter [6]. Often, natal dispersal distances between sexes are different, which helps to avoid inbreeding [7,8]. For example, in golden eagles (*Aquila chrysaetos*), females have been observed to have longer natal dispersal distances than males [9].

Despite the costs of dispersal in terms of time, energy, and predation risk, it must be beneficial to the individual. The main suggested benefits of dispersal include improved breeding conditions, inbreeding avoidance, and reduction in variance in breeding success through the bet-hedging strategy, where individuals increase their chances of successful breeding by alternating mates or nest sites in an unstable environment [10,11]. Especially, breeding success in a previous year has been suggested as a determinant of breeding dispersal [4,12]. In raptors, unsuccessful breeding has led to breeding dispersal, for example in burrowing owls (*Athene cunicularia*) [13]. However, this has not been observed for golden eagles or peregrine falcons (*Falco peregrinus*) [14,15]. Habitat quality variation is another important factor in determining breeding dispersal [1], and it has been found that individuals breeding in suboptimal territories are more prone to disperse than individuals breeding in good territories [16]. This type of behavior has been observed in, for example, Finnish female goshawks (*Accipiter gentilis*), which were dispersing to habitats less barren than their initial breeding site [17]. However, the role of habitat quality for breeding dispersal has been questioned, and other factors, such as mate or habitat loss, have been proposed [17]. Lastly, breeding dispersal may depend on individual characteristics, such as sex, age, and body condition [4], and interact with a high population density via increased competition over resources in habitats with low carrying capacity [18]. For example, in common kestrels (*Falco tinnunculus*), breeding dispersal distance varied between sexes, and was affected by prey abundance [19].

Turnover (i.e., change) of one or both breeding adults in a territory may be explained by breeding dispersal, but also by mortality or divorce [20,21]. Divorce is rare in raptors, but some cases have been observed in golden eagles where the female has found a new mate on another territory after an unsuccessful breeding [22]. Alternatively, the turnover event may be caused by death of one of the breeding individuals, which forces the other individual to find a new mate, and possibly a new territory.

Turnover is challenging to study, as it requires consistent data from multiple years. The data can be collected from observations of identifiable individuals or by using genetic tools. Genetic identification for turnover studies in raptors has been used for gyrfalcons [20], eastern imperial eagles (*Aquila heliaca*) [21], peregrine falcons [15], and goshawks [23]. Genotyping has been performed, mainly using shed adult feathers [20,21,23], but also using chick feathers or blood [15,20], which provide a more reliable source of DNA, as invasively collected samples from chicks have lower error rates than shed adult feathers [24].

In this study, we estimated the genetic diversity and population structure of golden eagles in Finland, using both nuclear microsatellites and mitochondrial control region sequences, to understand genetic parameters of the population. Then, for the first time in golden eagles, we calculated the annual and overall turnover rates for the Finnish population by genotyping chick feathers collected during a 15-year period. With the addition of breeding data, we looked for a relationship between breeding success and turnover, and between breeding success and nest site change within a territory, thereby testing the hypothesis that an unsuccessful breeding leads to turnover or nest site change. Finally, we studied natal dispersal for the individuals that were sampled first as chicks and later as adults.

2. Materials and Methods

2.1. The Study Species

Golden eagles are listed as vulnerable in Finland [25], but globally the species is classified as Least Concern by the IUCN [26]. The species was distributed throughout Finland excluding the Åland Islands up until the 1800s. However, golden eagles were hunted extensively during the 19th and 20th centuries, which severely decreased the population size across the whole country and led to the disappearance of the species in Southern Finland. In 1969, the species was fully protected, and the population began to recover in the 1970s [27]. At present, the distribution centralizes in the northern part of the country, with 90% of the pairs breeding in the north of the Oulujoki river, in the reindeer herding area [28]. Currently, there are 351–482 estimated breeding pairs in Finland [29]. The monitoring of golden eagles began in 1958 and continues to the present day by the

Parks & Forests Finland (Metsähallitus) [27]. Many, but not all, of the known nest sites are visited yearly (92% on average during 1971–2021), and the chicks are ringed, when possible. The adult birds are generally non-migratory in Finland, and their average territory size is around 150 km^2 with one to three nests per territory [30]. The nests are typically built on old pine or spruce trees [29]. During the last decade, the breeding success throughout Finland was the lowest in 2018, with only 92 successful nests and 108 chicks, while 2019 had the highest breeding success in the history of the monitoring program, with 175 successful nests and 211 chicks [31]. Factors that seemed to contribute to breeding success included prey abundance and weather conditions [31,32]. Main threats to golden eagles in Finland are illegal killings, habitat destruction, collisions with vehicles, powerlines, and wind turbines, as well as accumulation of toxins such as lead [33].

Golden eagles can live up to 34 years in the wild [34]. Mortality is low in adults (survival rate estimates range from 0.91 to 0.97), but very high in pre-adults (survival rates range from 0.23 to 0.50) [22,35]. Golden eagles start breeding at four to five years old and tend to mate for life. However, if one partner gets severely injured or dies, the other will find a new mate [22]. The species has a strong nest-site fidelity, and during the non-breeding years the pair maintain their bond by repairing nests or building new ones [36]. Territories often have several alternative nests, and, while some pairs may use a single nest throughout their breeding, others may alternate between nests more frequently [14,37]. Nest site changes may result from disturbance, persecution, or parasites, or the pair may use a different nest to maintain ownership of their territory [22]. In territories where good nest sites are rare, the pair will reuse nests more often, while in territories with multiple suitable alternative nests the pair may alternate between them more often [22,38].

2.2. *Sampling and Laboratory Analyses*

Feather samples were collected by volunteers of Metsähallitus during the ringing of chicks in 2006–2020 (Figure 1). Adults' shed feathers were collected from the nest and the surrounding area, while feathers from chicks were sampled from the birds. In total, 2215 samples were collected for DNA analyses (chicks N = 935 and adults N = 1280; Figure 1a,b). For each sample, information of the nest ID, location, territory, date, and chick ring number (when ringed) were recorded. Metsähallitus also provided data on breeding success within all territories including those where no feathers were collected.

Genomic DNA was extracted from the quill end of the feathers using QuickExtract™ solution (Epicentre) following the manufacturer's protocol. DNA concentration was measured with NanoDrop (Thermo Scientific, Waltham, MA, USA). Twelve polymorphic microsatellite loci were chosen for genotyping [39] (Table S1). The PCR reactions were prepared in a total volume of 6 µL using Qiagen Multiplex PCR kit (Qiagen, Hilden, Germany), and contained 3 µL of multiplex master mix, 0.6 µL of primer mix (Table S1), and 2.4 µL of genomic DNA. The concentration of the DNA varied a lot, depending on the quality of the sample, from ~10 to 1800 ng/µL. PCR profile started with initial denaturation at 95 °C for 15 min, followed by 35 cycles at 94 °C for 30 s, 60 °C for 90 s, and 72 °C for 60 s, with a final elongation at 60 °C for 30 min. The amplified 1–2 µL of PCR products were genotyped using GeneScan 500™ LIZ (Thermo Scientific) and formamide with ABI Prism 3730 Genetic Analyser (Applied Biosystems, Waltham, MA, USA).

A fragment of a mitochondrial DNA (mtDNA) control region of 88 golden eagle chicks was amplified and sequenced using primers modGOEA_CR1L (5′-CCC CCG TAT GTA TTA TTG TA-3′) [40] and GOEA_CR595H (5′-GCA AGG TCG TAG GAC TAA CC-3′) [41]. The chicks were selected to cover the whole sampling area, with one chick sampled per territory (Figure 1c). PCR reactions were carried out in a total volume of 10 µL. The final concentrations were as follows: 1X BIOTOOLS buffer, 4 mM $MgCl_2$, 0.2 mM of each dNTP, 0.7 µM of each primer, 1 mg of bovine serum albumin (BSA), and 0.1 units of BIOTOOLS polymerase (BIOTOOLS, Spain), and 1 µL (c. 100–1700 ng) of genomic DNA was used as a template. The PCR profile was 94 °C for 5 min followed by 45 cycles at 94 °C for 30 s, 57 °C for 60 s, and 72 °C for 60 s, and a final elongation at 72 °C for 5 min. Amplified PCR products were purified using the ExoI/FastAP (Thermo Scientific) method. Sequencing reactions were prepared with BigDye™ Terminator v3.1 kit (Thermo Scientific) in a total volume of 10 μL, which contained 1.5 μL of sequencing buffer (5X), 1 μL of ready reaction mix, 2 mM of a primer, and 5 μL of the purified PCR product. The PCR profile was 96 °C for 1 min, followed by 28 cycles at 96 °C for 10 s, 53 °C for 5 s, and 60 °C for 4 min. Sequencing reactions were purified using Sephadex filtration (Sigma-Aldrich, Taufkirchen, Germany) and sequenced with an ABI Prism 3730 Genetic Analyzer.

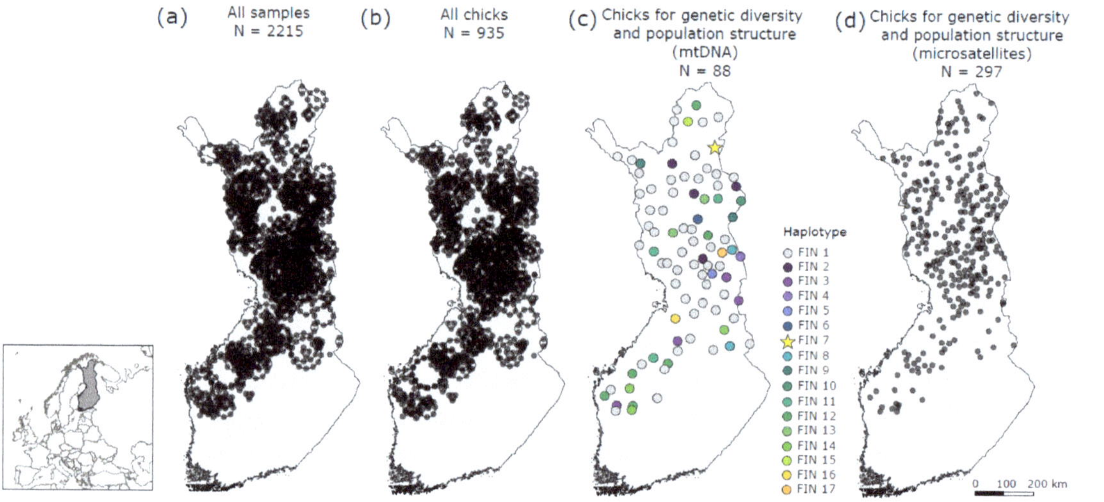

Figure 1. Sampling locations of individuals used in different analyses. (**a**) All samples. (**b**) All chick samples. (**c**) Chick samples used for genetic diversity and population structure analyses using mtDNA, with color codes for the detected haplotypes, indicating lack of population genetic structure. The Mediterranean haplotype (FIN7) is marked with the yellow star. (**d**) Chick samples used for genetic diversity and population structure analyses using microsatellites. Color intensity in panels (**a**,**b**) indicate sampling density, i.e., the darker the color, the more samples there are from the location, and multiple samples from a single location are arranged in rings.

Sexes of ten chicks that were used for studying natal dispersal were determined using the amplification refractory mutation system (ARMS) using three primers: P2 (5′-TCT GCA TCG CTA AAT CCT TT-3′) [42], NP (5′-GAG AAA CTG TGC AAA ACA G-3′) [43], and MP (5-AGT CAC TAT CAG ATC CGG AA-3′) [43]. The PCR reactions were carried out in a total volume of 10 μL. The final concentrations were as follows: 1X BIOTOOLS buffer, 2.5 mM $MgCl_2$, 0.2 mM of each dNTP, 0.8 μM of P2 primer, 0.4 μM of NP primer, 0.4 μM of MP primer, and 0.2 units of BIOTOOLS polymerase, and 1 μL (c. 100–1700 ng) of genomic DNA as a template. The PCR profile started with initial denaturation at 94 °C for 3 min, followed by 32 cycles of a PCR touchdown at 56–45 °C (94 °C for 30 s, 56–45 °C for 45 s, and 72 °C for 45 s, with two degrees intervals at 56–46 °C, each 2 cycles, and 20 cycles at 45 °C), with final elongation at 72 °C for 5 min. The PCR products were checked by 1.5% agarose gel electrophoresis at 145 V for 30 min and visualized under UV light. A single band was amplified in males, and two bands in females.

2.3. Genotyping and Sequencing Quality

The alleles were scored with GeneMapper v.5.0 software (Applied Biosystems). To assess genotyping errors, 372 samples (corresponding to 17% of the data) were duplicated, and, of those samples, 155 (corresponding to 7% of the data) were triplicated (see Table S2 for datasets of the analyses). The error rate was calculated per loci using Microsat_errcalc [44]. The program also estimated rates of allele dropouts, false alleles, and other errors such as mistypings. Allele dropouts and false alleles were estimated only for samples that had been triplicated as they had more reliable consensus genotypes.

Consensus genotypes were formed from the replicated samples. When it was not possible to determine the consensus, the genotype was called missing. Cervus [45] was used to estimate the observed and expected heterozygosities, polymorphic information content (PIC), and the average non-exclusion probabilities for identity and sibling identity for each locus. PIC is a measure of the degree of genetic diversity of a population with a selected set of loci, and it is used to identify markers that are particularly informative for studying the genetic structure and diversity of a population. Non-exclusion probability for identity refers to the likelihood that the genotypes of two unrelated individuals do not differ by chance. Similarly, non-exclusion probability for sibling identity is the likelihood that the genotypes of two siblings do not differ by chance. Thus, the lower the non-exclusion probability values are, the better the chosen set of loci can distinguish between individuals. Presence of null alleles was checked using Micro-Checker v.2.2.3 [46].

The primer for mtDNA sequencing was chosen by sequencing nine individuals with both forward (modGOEA_CR1L) and reverse (GOEA_CR595H) primers and visually comparing the quality of the obtained sequences in BioEdit v. 7.2.5 [47]. Based on this, the primer GOEA_CR595H was used for sequencing the rest of the samples.

2.4. Data Analysis

2.4.1. Genetic Diversity and Population Genetic Structure

Adult samples and individuals with over 25% missing data were removed from the analyses of genetic diversity and population structure. The data were resampled to only include one chick per territory to avoid sampling of relatives. The final dataset for estimating genetic diversity and population structure analyses included 297 golden eagle chicks (Figure 1d and Table S2). The analyses were carried out separately for each year and the sample sizes for each year were as follows: year 2006 (N = 9), 2007 (N = 30), 2008 (N = 16), 2009 (N = 13), 2010 (N = 8), 2011 (N = 21), 2012 (N = 28), 2013 (N = 30), 2014 (N = 34), 2015 (N = 30), 2016 (N = 17), 2017 (N = 13), 2018 (N = 12), 2019 (N = 25), and 2020 (N = 11).

Number of alleles (A), allelic richness (AR), and inbreeding coefficient (F_{IS}) were estimated using Fstat v. 2.9.4 [48]. The observed (H_O) and expected heterozygosity (H_E) were estimated using Arlequin v. 3.5 [49]. Linkage disequilibrium was evaluated on Genepop v. 4.7 [50].

Population structure was estimated using the Bayesian model-based clustering program Structure v. 2.3.4 [51]. Structure uses Markov chain Monte Carlo (MCMC) simulations to estimate the number of genetically distinct clusters (K). Using the admixture model, the program was run for 100,000 MCMC replicates with a burn-in of 10 000 for ten iterations, with K set to one to five, and without the location information (i.e., no locprior). The optimal K value was chosen based on the standard log probability test (LnP(K)) using Structure Selector [52]. LnP(K) was used because it allows for K = 1, unlike Delta (K) [53], and because it does not require pre-defined groups, like the Puechmaille's Optimal K [54]. The results were visualized on Pophelper v. 1.0.10 [55].

The mtDNA sequences were manually edited and aligned following ClustalW Multiple Alignment [56] with 1000 bootstrap replications in BioEdit. The number of haplotypes (H), haplotype ($\hat{h}$), and nucleotide (π) diversities, and the mutation parameter theta (ϑ) from the number of segregating sites (S), were calculated using the program DnaSP v. 6.12 [57] for the total dataset. A median joining haplotype network [58] was built in PopArt [59] and finalized in Inkscape v. 1.1 [60] to inspect visually for possible geographic clustering of haplotypes. For this, the individuals were divided into northern and southern groups, based on location either within the reindeer herding area or outside of it. The haplotype and trait files were created in RStudio v. 1.4 [61] with packages "pegas" [62] and "ape" [63].

2.4.2. Turnover

Relatedness (r) between all chicks (N = 935, Table 1), was estimated with the program ML-relate, which calculates maximum likelihood estimates of relatedness and relationship considering null alleles [64]. Relatedness values were estimated between chicks from different years within the same territories. The data included 120 known pairs of siblings. When there was a 40% drop in relatedness between years within a territory, it was interpreted as a change in one of the parents, and if there was more than 85% drop it was interpreted as a change of both parents. Drops of 40% instead of 50% and 85% instead of 100% were chosen to consider possible genotyping errors and missing data [15]. The turnover rates were calculated by dividing the number of turnover events (i.e., change of one or both parents within a territory) by the number of all comparisons between years within territories (i.e., all possible turnover events). The statistical difference between 'known brood relatedness' and 'within territory relatedness' was assessed using the Wilcoxon signed-rank test in RStudio. Similarly, the difference between 'within territory relatedness' and 'relatedness in territories where turnover occurred' was tested.

Table 1. Number of sampled chicks, nests, and territories during years 2006–2020 where samples were collected. Same territories were sampled in different years, and different nests from the same territories were sampled in different years, resulting in a different total number of nests and territories.

Year	Chicks	Nests	Territories
2006	25	24	24
2007	109	84	84
2008	58	53	53
2009	39	35	35
2010	30	29	29
2011	75	61	61
2012	80	70	70
2013	81	72	72
2014	103	86	86
2015	101	91	91
2016	36	33	33
2017	42	39	39
2018	35	30	30
2019	65	58	58
2020	56	47	47
Total	935	455	297

To test whether breeding success of a previous year affected turnover, a chi-squared test was performed in RStudio. Similarly, the dependency between breeding success and nest site change was tested using the chi-squared test. The null hypotheses for the tests were "turnover is independent of breeding success" and "change in a nest site is independent of breeding success".

2.4.3. Natal Dispersal

To estimate natal dispersal, identity analysis was performed in Cervus, with a minimum of ten matching loci allowing for one mismatch and using the total dataset of chicks and adults. Two matching genotypes were classified as cases of natal dispersal when one genotype was from a chick and another one was from an adult of four and more years apart, which is the age that golden eagles reach sexual maturity. The distance between the two nests was then measured using QGIS 3.10 [65]. When a chick genotype matched to an adult from several different years, only the distance to the first detected adult was measured.

3. Results

3.1. Genotyping Quality

In total, 935 chick and 1278 adult samples were genotyped. The genotyping error rate for the total dataset was 0.0621, and varied from 0.011 in locus NVHfr206 to 0.089 in locus Aa04 (Table S3). The overall amplification success was 84%. However, since genetic diversity, population structure, and turnover were studied with the chick dataset, removal of poorly amplified adult genotypes reduced the error rate and increased the amplification success. Thus, the error rate for all loci in the chick dataset was 0.024, ranging from 0.000 to 0.089 with 41 samples replicated (Table S4). The amplification success of chick genotypes was 95%, ranging from 68% to 100% among loci. Locus Aa39 produced scorable alleles in only c.30% of cases, and, therefore, it was removed from further analyses.

In the chick dataset, null alleles were found in loci Aa35, Aa36, and NVHfr124. When dividing the data into yearly datasets, no consistent pattern was observed, suggesting that in some years null alleles could have resulted from an undetected subpopulation structure. Therefore, all loci were kept for the downstream analyses. The combined non-exclusion probability of identity was 2.185×10^{-9}, and the combined non-exclusion probability of sibling identity was 0.0002 for the whole dataset. In the chick dataset, the combined non-exclusion probability of identity was 1×10^{-8}, and the combined non-exclusion probability of sibling identity was 0.0005.

3.2. Genetic Diversity and Population Genetic Structure

Genetic diversity estimates for the chick dataset (N = 297) are presented in Table S4. The mean observed heterozygosity was 0.567, ranging from 0.035 to 0.752 per locus. The expected heterozygosity varied from 0.034 to 0.813 per locus, with an average of 0.585. The mean number of alleles was 7.9 and varied from 3 to 16 among loci. The allelic richness varied from 1.34 to 5.88 per locus, with an average of 3.87. The inbreeding coefficient varied from −0.058 to 0.109 between loci, with an average of 0.033.

Genetic diversity estimates measured yearly over all loci for the chick dataset are presented in Table 2. The observed heterozygosity was the highest in 2020 (0.633) and lowest in 2014 (0.551). The expected heterozygosity was the lowest in 2008 and 2014 (0.551), and the highest in 2020 (0.631). The highest number of alleles was in 2007 (5.82) and the lowest in 2010 (3.73); allelic richness on the other hand was the lowest in 2008 (3.57) and the highest in 2020 (4.16). No significant deviations from the Hardy–Weinberg equilibrium were observed in the yearly datasets.

Table 2. Nuclear genetic diversity estimates of 11 microsatellite loci for yearly resampled chick datasets. Number of samples, observed (H_O) and expected (H_E) heterozygosities, number of alleles (A) and allelic richness (AR) based on a minimum of seven diploid individuals, and inbreeding coefficients (F_{IS}).

Year	N	H_O	H_E	A	AR	F_{IS}
2006	9	0.586	0.584	4.27	3.95	0.032
2007	30	0.593	0.593	5.82	4.04	0.018
2008	16	0.552	0.551	4.64	3.57	0.061
2009	13	0.618	0.618	4.73	3.94	0.009
2010	8	0.574	0.573	3.73	3.63	0.007
2011	21	0.580	0.579	5.09	3.79	0.055
2012	28	0.581	0.580	5.73	3.80	0.034
2013	31	0.594	0.592	5.18	3.76	−0.045
2014	34	0.551	0.551	5.45	3.63	0.053
2015	30	0.603	0.603	5.45	3.92	0.015
2016	17	0.615	0.613	5.55	4.09	0.096
2017	13	0.609	0.707	4.64	3.96	0.089
2018	12	0.569	0.569	4.64	3.88	0.008
2019	25	0.583	0.582	5.73	3.86	0.046
2020	11	0.633	0.631	4.73	4.16	0.063

The most likely number of clusters (K) for the Finnish golden eagle population was one (mean LnP (K = 1) = −7805.87, and mean LnP (K = 2) = −8182.81). The likelihood decreased further with a higher K (Figure S1).

For the mtDNA, a 393 bp alignment from the 88 sequenced individuals was obtained. This alignment had 18 segregating sites, forming 17 haplotypes (GenBank accession numbers: OQ679875-OQ679891; Table S5). Of these haplotypes, 16 belonged to the Holarctic lineage and one to the Mediterranean lineage, as defined by [40]. The haplotype network showed no clustering according to geographic locations; however, only five haplotypes were shared between the northern and the southern groups (Figure S2). The total haplotype diversity was 0.617, nucleotide diversity 0.0034, and theta 0.0091.

3.3. Turnover

The mean sibling relatedness was 0.496 for 120 pairs of known full siblings. To give an understanding of how the relatedness values were distributed, there were 22 cases where relatedness was lower than 0.3, and 22 cases where it was over 0.7. Chicks from the nests with one parent turnover (cut-off of 40%) had a mean relatedness of 0.298, and from the nests where both parents had changed (cut-off of 85%) had a mean relatedness of 0.074. The mean relatedness within territories was 0.326, ranging from 0 to 0.721, while the mean relatedness for all territories where turnover occurred was 0.170. There was a significant difference between the 'known brood relatedness' and 'within territory relatedness' (V = 2814, p = 0.0327), and between the 'within territory relatedness' and 'relatedness in territories where turnover had occurred' (i.e., 'territories with below the 40% cut-off', V = 2775, $p = 7.893 \times 10^{-14}$).

In total, 201 territories had more than one year of breeding during the sampling period, and, therefore, were included in the turnover calculations. The mean number of chicks per territory across years was four, and the mean number of years when samples were collected was 3.5. The highest number of sampled chicks per territory was 15, and the highest number of sampled years in a territory was 11. Nine territories were excluded due to uncertainties in individual and territory IDs.

The overall population turnover rate was 23% in all the years combined. The yearly rates varied from 0% in 2007, 2009, and 2010 to 38% in 2020 (Figure 2). The total number of turnover events in the 15-year dataset was 101, with the highest number in 2015 when 20 turnover events were observed. In 2015, half of the events were with a change in both parents and half with one parent only (Figure 3). The lowest numbers of turnover events were in the years 2007–2011 and in 2016–2018. Since 2006 was the starting year of this study, it could not be compared with a previous year to detect any turnover events. In total, there were 57 changes of one parent and 44 changes of both parents. In most years, the frequency of turnover events involving one parent and both parents were similar, except for 2014, which had 4 turnover events of both parents and 11 of one parent.

The turnover was dependent on the previous year's breeding success (χ^2 = 21.04, df = 1, and p = 0.000004). The number of turnover events after a successful breeding was 33, and 68 after an unsuccessful breeding. No turnover occurred after a successful breeding in 206 cases and after an unsuccessful breeding 146 times (Figure 4a). The nest site change within a territory was dependent on the previous year's breeding success as well (χ^2 = 91.67, df = 1, and p = 0.000000). The nest change occurred 145 times after an unsuccessful breeding and 54 times after a successful breeding. The breeding pair remained in the same nest after a successful breeding year 185 times and 69 times after an unsuccessful breeding (Figure 4b).

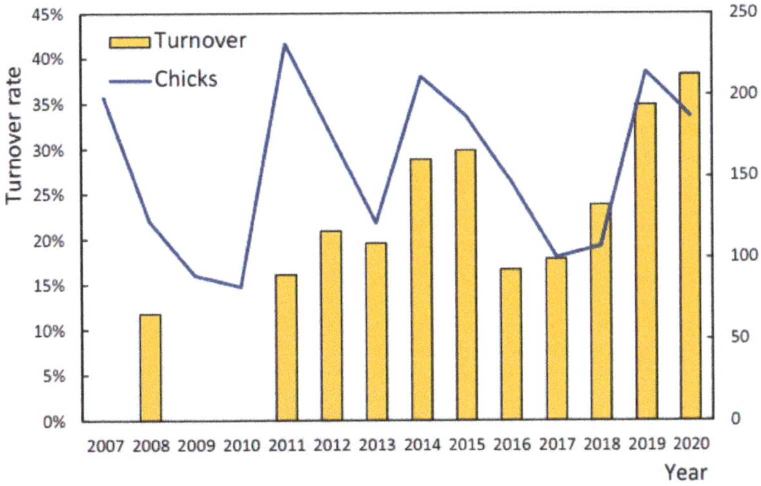

Figure 2. Turnover rate of parents each year, estimated from 11 microsatellite loci, as bars, and the actual number of chicks per year as a line.

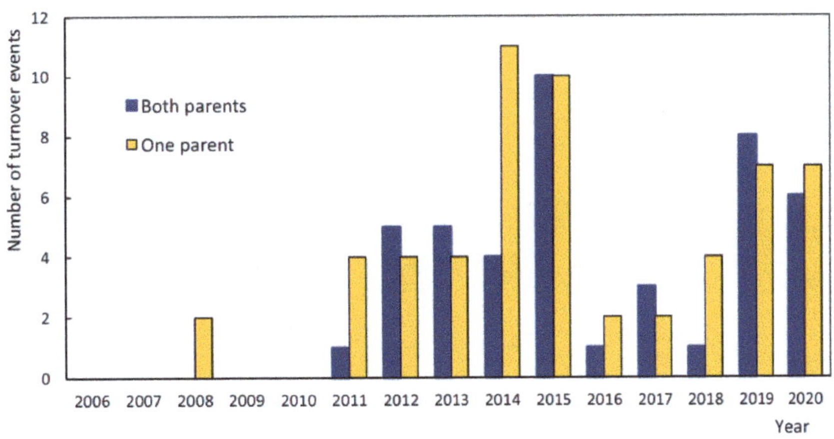

Figure 3. Number of turnover events estimated from 11 microsatellite loci of both (blue bars) or one parent (yellow bars) each year from 2006 to 2020.

3.4. Natal Dispersal

There were ten cases where an individual had been sampled as a chick and four to nine years later as an adult (Figure 5). Eight of the ten birds were females and two were males. The distances between natal and first detected breeding nest sites varied from 15.0 to 372.2 km, with a mean of 110.2 km and a median of 98.1 km. Nine of the re-sampled individuals stayed in the region where they were born, forming two geographical clusters: Northern Finland (N = 5; of which four were females and one was a male) and Southern Finland (N = 4; of which three were females and one was a male). Only one individual changed regions, which was a female born in the north moving to the south for breeding. When comparing the distances between these two clusters, the mean dispersal in the northern region was 96.2 km, with a median of 118.1 km, ranging from 15 km to 133 km, while in the southern region it was 62.2 km on average, with a median of 51.1 km, ranging from 40 km to 106 km. The natal dispersal distance in females (N = 8) was 130 km on average, with a median of 112.2 km, ranging from 40.5 km to 372.2 km, and the distances in males (N = 2) ranged from 15.0 km to 46.8 km. The sample sizes were too small to test statistically whether there was a significant difference in the distances between the two regions or sexes.

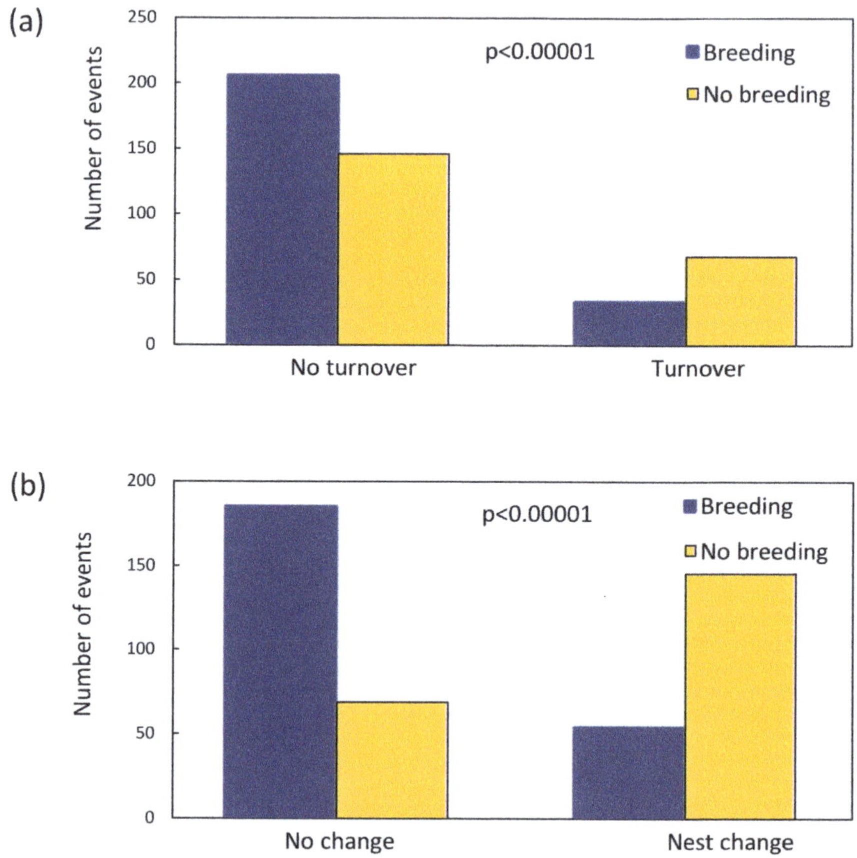

Figure 4. (**a**) Number of breeding events with and without turnover following successful (blue bars; breeding) and unsuccessful (yellow bars; no breeding) breeding attempts in the previous year. (**b**) Number of nest site changes within a territory when there was a successful (blue bars; breeding) or an unsuccessful (yellow bars; no breeding) breeding in the previous year, and the number of occasions when the pair used the same nest after a successful or an unsuccessful breeding year.

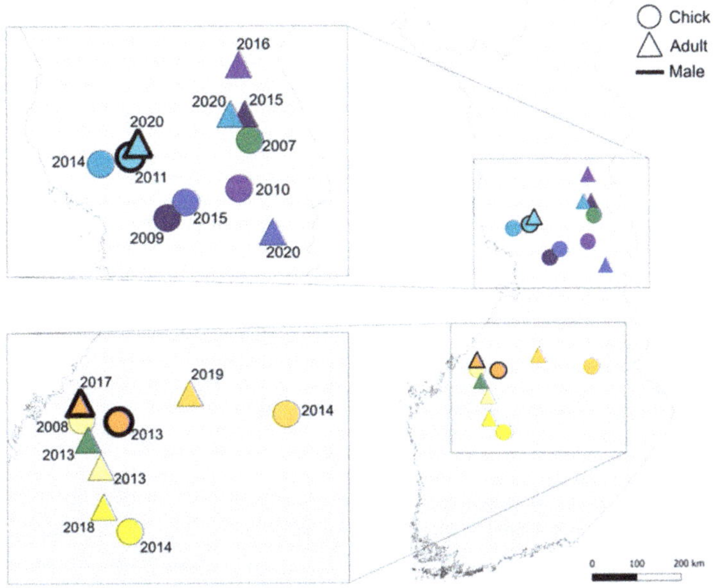

Figure 5. Locations of ten golden eagle chick samples (circles) that were later identified as adults (triangles) using 11 microsatellite loci. Each individual has a unique color code, which is maintained for chick and adult ages. Years of sampling are indicated near the sampling points. The two males are indicated by symbols with a bold outline. Individuals from the northern group are represented by blue color shades, and from the southern group by yellow color shades. One individual that was born in Northern Finland and later was breeding in Southern Finland is colored in green.

4. Discussion

4.1. Genetic Diversity and Population Genetic Structure

The level of genetic diversity of Finnish golden eagles estimated from the nuclear loci remained quite stable throughout the study years. When compared with previously reported estimates for the Estonian–Finnish population [66], the Finnish population in this study had similar, yet slightly lower, observed and expected heterozygosities (H_O = 0.57 and 0.60, and H_E = 0.59 and 0.60 for Finland and Estonia–Finland, respectively). However, the Estonian–Finnish population in the previous study [66] had a substantially smaller sample size (N = 24) and consisted mostly of individuals from Estonia (N = 20), while here the estimates were from 297 Finnish golden eagles. The nuclear genetic diversity of golden eagles from Finland is at the same level as those reported for golden eagles from North America (H_O = 0.52–0.62, and H_E = 0.48–0.55) [41,67], British Isles (H_O = 0.43–0.51, and H_E = 0.48–0.51) [68,69], and Japan (H_O = 0.52–0.59, and H_E = 0.55–0.56) [70].

Meanwhile, mitochondrial genetic diversity was higher in Finland ($\hat{h}$ = 0.62, and π = 0.0034) than previously reported for Fennoscandia ($\hat{h}$ = 0.41, and π = 0.0010) [40]. In the previous study [40], the sample size was much smaller compared with this study (N = 38 and 88), which could have affected the estimates, so comparison of results should be made with caution. In addition, the Fennoscandian population in the previous study was mostly represented by individuals from Norway (N = 34), suggesting that the Finnish population has a higher genetic diversity than the Norwegian population, which may be because the latter is at the Eurasian northwesternmost edge of the eagle's distribution. However, on a wider geographical scale, diversity values for mitochondrial control region sequences vary a lot (for example, in North America: $\hat{h}$ = 0.49–0.81 and π = 0.0013–0.0030 [41,67,71]; and in

Eurasia: $\hat{h}$ = 0.41–0.93 and π = 0.0010–0.0090 [40,70]), likely depending much on the sample size and sequence length.

No population genetic structure in Finland was found with the used markers, in accordance with previous studies of the Holarctic region [40,66]; however, only 5 of 17 mitochondrial haplotypes were shared between northern and southern parts of the country. Surprisingly, one individual carrying mitochondrial DNA of the Mediterranean lineage was found in northern Finland, indicating that long-distance dispersal of golden eagles from south to north can occur.

4.2. Turnover Rate

While golden eagles are considered to be monogamous and mate for life, and to be highly territorial, the turnover rate of 23% suggests that this idea may not be that straightforward. The cost of turnover could have been expected to be high for such a species. Finding new territories can be challenging and lead to intraspecific aggression in crowded regions, while maintaining a territory is less costly [16]. However, the benefits of leaving a territory may outweigh these costs. For example, if a habitat change leads to increased breeding success, the pair may leave the territory to increase their chances of breeding [1]. Similarly, monogamy has several advantages, such as paternity assurance and reduced mate competition, which may at the same time decrease divorce rates [72]. However, if the breeding mate shows a lack of investment, or if the breeding fails, divorces may occur [22,72]. Turnover may also be a forced behavior, for instance due to habitat loss caused by urban and agricultural land expansion, or due to the death of a mate [73,74]. Especially, older individuals with settled territories, successful breeders, and individuals inhabiting high-quality territories are more likely to remain within a territory [1] but might be forced to change territories due to external factors.

In golden eagles, the number of turnover events of one parent was slightly higher than that of both parents. It is not known whether the other member of the pair had died or if the birds changed partners. However, as mortality for adult birds is low [33], it is unlikely that mortality alone would have caused all the one parent turnover events. When both parents were changed in a territory, it was possibly a result of breeding dispersal, or, if the male died, the female might have left the territory. While mortality is low for adult golden eagles, there are some factors that may contribute to higher-than-expected mortality in certain territories. Probable mortality causes for Finnish golden eagles include intraspecific aggression, persecution, lead poisoning, and collisions with cars, trains, powerlines, and wind turbines [33]. In particular, collisions may prove to be a higher risk in territories near powerlines or windfarms. On the other hand, persecution that used to be higher in specific areas, such as in the reindeer herding area (Northern Finland), where golden eagles prey on reindeer calves, has diminished and should not be concentrated to any specific area anymore due to the incentives paid by the government [75].

Divorces in golden eagles have been observed using radiotracking methods, when females changed a mate and a territory after an unsuccessful breeding with previous mates. The males remained in the territory and attempted to breed with a new female but the breeding failed again [22]. Genetic data has not previously been used to detect divorce in golden eagles, but, based on the high number of turnover events of one parent found in this study, it is likely that some divorces do occur. Moreover, for many bird species that have been considered monogamous, genetic studies have proven this to not always be the case [76,77].

In addition to mortality and divorce, extra-pair copulation may have resulted in the apparent one parent turnover events. Extra-pair copulation, specifically extra-pair paternity, is common in birds [78–80]. There have been no observations of extra-pair paternity in golden eagles, although observations of three eagles in one territory have been made in Scotland, Sweden, and Norway ([81–83] cited in 22). These records have been made based on visual observations, and it was not certain in all cases whether the third eagle was a male or a female. The three eagles participated in the incubation and prey provision for the chick. In Scotland, an additional male was observed displaying with the resident female while the resident male was present ([82] cited in 22). Based on the relatedness values between suggested full siblings in our dataset, it is possible that extra-pair paternity may have occurred. Further parentage analysis using the adult samples might provide more information on the issue.

Another contributing factor to turnover could be unclear territory boundaries. In Scotland, golden eagles were found to take over nests of neighboring territories when they were unoccupied, or absorb parts of other territories [84]. Whether this occurs in Finland is unknown. The breeding adult eagles in Finland are resident and highly territorial, so the territories might be more rigid and less dynamic. Still, it is possible that some territories are more complex. Moreover, the individual identity of the nest or territory determined by the data collector may not correspond to reality, especially when nests are close together. Clarification of the boundaries will require very intensive and continuous monitoring of all territories.

The observed turnover rates varied yearly, with the highest rates found in the later years. It might seem that the turnover rate increases over time (Figure 2), but, more likely, it is merely a result of the data structure and sampling. In many cases, the samples were from a few years apart, either because there was no breeding, no samples collected, or the sample had been excluded from the analyses. Therefore, it is possible that the later years showed a higher turnover rate since there was a higher chance of observing it. Thus, the yearly variation in turnover is most likely an artifact of the data structure and sampling and should not be used to describe the population in those years.

Turnover rate has not been estimated for golden eagles previously, which makes these results valuable, but also prevents the evaluation of results in relation to previous research. Fortunately, turnover has been estimated for some other raptor species, such as the gyrfalcon, with a 20% turnover rate [20], the eastern imperial eagle, with a 28–36% turnover rate [21], and the peregrine falcon, with a 21.7% turnover rate [15]. It appears that the estimated turnover rate for golden eagles in Finland conforms with the estimates in the other raptors. All these species have similar breeding behavior: they are known to be monogamous and to return to the same territory with alternative nests to breed, which allows for some comparisons. Moreover, the estimated turnover rate in peregrine falcons was performed using a similar method as in this study, allowing for a more robust comparison of results.

4.3. *Turnover and Nest Change as a Result of Unsuccessful Breeding*

Another significant finding of this study was that turnover and change of nest site within a territory seemed to occur more often after an unsuccessful breeding. This has been previously hypothesized [14,85], but has never received statistical support in golden eagles [22,86]. In the case of one parent turnover after an unsuccessful breeding, divorce may have occurred. Because of the differences in costs of breeding dispersal between sexes, it is likely that the female will leave the territory to find a new mate, while the male remains in the territory [72]. However, it is not possible to determine which parent left the territory with our data. One parent turnover after a successful breeding could be caused by mortality, as there is no obvious advantage in switching mates after a successful breeding because the benefits of monogamy are high [72], or by extra-pair copulations.

In cases of no turnover, the pairs appeared to remain together and within the same territory more often after successful breeding. Lack of turnover suggests that the territory is highly suitable for breeding, which could result from high prey abundance, no disturbances, or/and low competition. One of the main factors affecting breeding success is prey availability [87]. Some prey species of golden eagles fluctuate yearly, leading to poorer breeding years when prey is scarce, which might cause the pair to leave their territory to find better breeding and hunting grounds. The most successful breeding year during the study period was 2019, and it was thought to result from increased grouse populations and high number of hares [31]. In the United States, breeding success of golden eagles has been found to correlate with jackrabbit (*Lepus californicus*) abundance [88]. In contrast, an earlier Finnish study found no effect of prey availability on territory occupancy in golden eagles [89], but there were no significant fluctuations in prey abundance during the study period. Similarly, no relationship between breeding success and food supply was found in Scotland, possibly also due to a short study period [90]. Studying correlations between prey abundance and breeding success requires detailed data of both the prey abundance and breeding success. In Finland, the sizes of grouse populations are estimated yearly [91], providing an opportunity for future research to study the relationship between grouse species abundance and the breeding success of golden eagles.

Weather conditions during the breeding season can also affect breeding success. Golden eagles are especially sensitive to poor weather conditions during incubation, which may lead to failed nesting [22,90]. In Finland, long cold and rainy springs have possibly reduced breeding success in some years and areas [32]. However, the relationship between weather and breeding success is complicated. Poor weather conditions may also affect prey abundance, which leads to lower breeding success, rather than directly affecting golden eagle breeding performance. Still, extreme weather conditions may directly lead to breeding failure, as observed in the United States, where a three-day blizzard resulted in 71% of nests containing chicks to fail [22].

In addition to turnover, unsuccessful breeding appeared to lead to a nest site change in the following year. Similar factors may contribute to the change in a nest site as to turnover. Since golden eagle territories are quite large, 151 km^2 on average [30], it is possible that different hunting grounds are available within the territory. Golden eagles may change nests to be closer to better hunting areas rather than completely abandoning their territory. In addition, there might be other factors within the territory that may affect choice of the nest site, such as destruction of the previous nest site or disturbance. However, a habitat assessment needs to be carried out to have a better understanding of this subject. Based on results from this study, it seems that the pairs favor a nest where they previously had a successful breeding.

4.4. Natal Dispersal

The natal dispersal of Finnish golden eagles was estimated for ten individuals. Despite the sample size being small, it provides new valuable information on golden eagle dispersal in Finland, because natal dispersal distances have not been estimated here previously.

The mean natal dispersal distance of Finnish golden eagles was 110.2 km, which was higher than the observed distance in southwestern United States (55.3 km) [9]. Individuals seemed to disperse further from their natal territories in Northern Finland, where population density is higher, than in Southern Finland. Similarly, estimates within the United States have also varied depending on geographical location, from a median of 46.6 km in arid southwestern to 64.5 km in less arid western United States [9]. Thus, in addition to density, differences in dispersal distances between regions could arise from differences in habitat quality [9]. Even though comparing the results should be carried out with caution due to small sample sizes (N = 16 in the USA, and N = 10 in Finland), in both studies the majority of samples was from females, making the results somewhat comparable. Previous studies have found that there might be a difference in natal dispersal distances between sexes, which could be a behavioral mechanism to avoid inbreeding [1]. In golden eagles, females seem to travel further than males [9], but in this study the sample sizes were too small (two males and eight females) to perform a statistical comparison between sexes.

5. Conclusions

This study provides novel information on the population dynamics of golden eagles in general, and on population genetics of golden eagles in Finland. We found no notable changes in genetic diversity during the 15-year study period and did not discover any population structure, which may be due to effective dispersal of the species. The turnover rate of 23% is high for a species who mates for life and has a strong nest-site fidelity.

These results suggest that there are multiple factors affecting breeding pairs, or golden eagles might not exhibit such extreme monogamy as believed. Based on the findings, golden eagles are prone to change territories, mates, or nests after an unsuccessful breeding. Further implications of this study might arise by performing more detailed analyses on the causes of turnover.

Supplementary Materials: The following supporting information can be downloaded at: https://www.mdpi.com/article/10.3390/d15040567/s1, Table S1: Microsatellite primers [92,93]; Table S2: Datasets used for different analyses with the number of samples used; Table S3: Characteristics of used microsatellite loci.in the total dataset; Table S4: Characteristics of used microsatellite loci in the chick dataset; Table S5: Mitochondrial haplotype information; Figure S1: Results from the structure analysis of Finnish golden eagles; Figure S2: Median-joining haplotype network of Finnish golden eagles.

Author Contributions: Conceptualization, S.P. and L.K.; methodology, A.K., E.K., S.P. and L.K.; software, A.K. and E.K.; validation, A.K., E.K., S.P. and L.K.; formal analysis, A.K. and E.K.; investigation, A.K., E.K., T.O., S.P. and L.K.; resources, T.O., S.P. and L.K.; data curation, T.O., E.K. and A.K.; writing—original draft preparation, A.K. and E.K.; writing—review & editing, A.K., E.K., T.O., S.P. and L.K.; visualization, A.K., E.K. and L.K.; supervision, E.K., S.P. and L.K.; project administration, S.P. and L.K.; funding acquisition, A.K. and L.K.; A.K. and E.K. equal contribution. All authors have read and agreed to the published version of the manuscript.

Funding: This research was funded by Oulun luonnonystäväin yhdistys and the University of Oulu.

Data Availability Statement: DNA sequences are available from GenBank under accession numbers OQ679875-OQ679891 and microsatellite data from the corresponding author on reasonable request. Since the golden eagle is a protected and still persecuted species, sensitive data on location of individuals and territories will not be released.

Acknowledgments: We thank Soile Alatalo, Hannele Parkkinen, Ian Russell-Jones, Ville Sannikka and Sujeet Singh for their assistance in the lab, and all the volunteer ringers of Metsähallitus for collecting samples.

Conflicts of Interest: The authors declare no conflict of interest.

References

1. Greenwood, P.J. Mating systems, philopatry and dispersal in birds and mammals. *Anim. Behav.* **1980**, *28*, 1140–1162. [CrossRef]
2. Linkhart, B.D.; Reynolds, R.T. Return rate, fidelity, and dispersal in a breeding population of Flammulated Owls (*Otus flammeolus*). *Auk* **2007**, *124*, 264–275. [CrossRef]
3. Cilimburg, A.B.; Lindberg, M.S.; Tewksburry, J.J.; Hejl, S.J. Effects of dispersal on survival probability of adult Yellow Warblers (*Dendroica petechia*). *Auk* **2002**, *119*, 778–789. [CrossRef]
4. Greenwood, P.J.; Harvey, P.H. The natal and breeding dispersal of birds. *Annu. Rev. Ecol. Syst.* **1982**, *13*, 1–21. [CrossRef]
5. Paradis, E.; Baillie, S.R.; Sutherland, W.J.; Gregory, R.D. Patterns of natal and breeding dispersal in birds. *J. Anim. Ecol.* **1998**, *67*, 518–536. [CrossRef]
6. Briggs, C.W.; Collopy, M.W.; Woodbridge, B. Correlates and fitness consequences of natal dispersal in Swainson's Hawks. *Condor* **2012**, *114*, 764–770. [CrossRef]
7. Clarke, A.L.; Sæther, B.-E.; Røskaft, E. Sex Biases in Avian Dispersal: A Reappraisal. *Oikos* **1997**, *79*, 429–438. [CrossRef]
8. Pasinelli, G.; Schiegg, K.; Walters, J.R. Genetic and environmental influences on natal dispersal distance in a resident bird species. *Am. Nat.* **2004**, *164*, 660–669. [CrossRef]
9. Murphy, R.K.; Stahlecker, D.W.; Millsap, B.A.; Jacobson, K.V.; Johnson, A.; Smith, C.S.; Tator, K.J.; Kruse, K.L. Natal dispersal distance of golden eagles in the southwestern United States. *J. Fish Wildl. Manag.* **2019**, *10*, 213–218. [CrossRef]
10. Doligez, B.; Pärt, T. Estimating fitness consequences of dispersal: A road to "know-where"? Non-random dispersal and the underestimation of dispersers' fitness. *J. Anim. Ecol.* **2008**, *77*, 1199–1211. [CrossRef]
11. Scheiner, S.M. Bet-hedging as a complex interaction among developmental instability, environmental heterogeneity, dispersal, and life-history strategy. *Ecol. Evol.* **2014**, *4*, 505–515. [CrossRef] [PubMed]
12. Dong, R.; Ye, X.; Zhong, L.; Li, X.; Li, M.; Wang, H.; Yu, X. Effects of breeding success, age and sex on breeding dispersal of a reintroduced population of the Crested Ibis (*Nipponia nippon*) in Ningshan County, China. *Avian Res.* **2018**, *9*, 1–7. [CrossRef]
13. Riding, C.S.; Belthoff, J.R. Breeding Dispersal by Burrowing Owls (*Athene cunicularia*) in Idaho. *J. Raptor Res.* **2018**, *52*, 143–157. [CrossRef]
14. Kochert, M.N.; Steenhof, K. Frequency of nest use by Golden Eagles in southwestern Idaho. *J. Raptor Res.* **2012**, *46*, 239–247. [CrossRef]

15. Ponnikas, S.; Ollila, T.; Kvist, L. Turnover and post-bottleneck genetic structure in a recovering population of Peregrine Falcons *Falco peregrinus*. *Ibis* **2017**, *159*, 311–323. [CrossRef]
16. Harts, A.M.F.; Jaatinen, K.; Kokko, H. Evolution of natal and breeding dispersal: When is a territory an asset worth protecting? *Behav. Ecol.* **2016**, *27*, 287–294. [CrossRef]
17. Otterbeck, A.; Lindén, A.; Gunko, R.; Ylinen, E.; Byholm, P. Female breeding dispersal to higher quality habitats in a philopatric top predator. *J. Ornithol.* **2022**, *163*, 3–92. [CrossRef]
18. Enfjäll, K.; Leimar, O. The evolution of dispersal—The importance of information about population density and habitat characteristics. *Oikos* **2009**, *118*, 291–299. [CrossRef]
19. Vasko, V.; Laaksonen, T.; Valkama, J.; Korpimäki, E. Breeding dispersal of Eurasian kestrels *Falco tinnunculus* under temporally fluctuating food abundance. *J. Avian Biol.* **2011**, *42*, 552–563. [CrossRef]
20. Booms, T.L.; Talbot, S.L.; Sage, G.K.; McCaffery, B.J.; McCracken, K.G.; Schempf, P.F. Nest-site fidelity and dispersal of Gyrfalcons estimated by noninvasive genetic sampling. *Condor* **2011**, *113*, 768–778. [CrossRef]
21. Vili, N.; Szabó, K.; Kovács, S.; Kabai, P.; Kalmár, L.; Horváth, M. High turnover rate revealed by non-invasive genetic analyses in an expanding Eastern Imperial Eagle population. *Acta Zool. Acad. Sci. Hung.* **2013**, *59*, 279–295.
22. Watson, J. *The Golden Eagle*, 2nd ed.; T & AD Poyser: London, UK, 2010.
23. Selås, V.; Kleven, O.; Steen, O.F. Female turnover rate differs between two Northern Goshawk *Accipiter gentilis* nesting areas, as revealed by DNA analysis of moulted feathers. *Ibis* **2017**, *159*, 554–566. [CrossRef]
24. Bush, K.L.; Vinsky, M.D.; Aldridge, C.L.; Paszkowski, C.A. A comparison of sample types varying in invasiveness for use in DNA sex determination in an endangered population of Greater Sage-Grouse (*Centrocercus uropihasianus*). *Conserv. Genet.* **2005**, *6*, 867–870. [CrossRef]
25. Hyvärinen, E.; Juslén, A.K.; Kemppainen, E.; Uddström, A.; Liukko, U.-M. *Suomen Lajien Uhanalaisuus 2019—Punainen Kirja: The 2019 Red List of Finnish Species*; Ympäristöministeriö & Suomen Ympäristökeskus: Helsinki, Finland, 2019.
26. BirdLife International Species Factsheet: Aquila chrysaetos. Available online: http://datazone.birdlife.org/species/factsheet/golden-eagle-aquila-chrysaetos (accessed on 6 March 2023).
27. Below, A. Suojelualueverkoston merkitys eräille nisäkäs-ja lintulajeille. In *Metsähallituksen Luonnonsuojelujulkaisuja, Sarja A, Nro 121*; Metsähallitus: Vantaa, Finland, 2000. (In Finnish)
28. Ollila, T.; Koskimies, P. Maakotkan ja muuttohaukan suojelutaso Suomessa. *Linnut Vuosik.* **2008**, *2007*, 8–17, (In Finnish with English Summary).
29. Ollila, T. *Raportti Maakotkan, Muuttohaukan, Tunturihaukan Sekä Oulun Ja Lapin Läänien Merikotkien Pesinnöistä Vuonna 2021*; Metsähallitus: Vantaa, Finland, 2021. (In Finnish)
30. Ollila, T. The monitoring project of the Golden Eagle *Aquila chrysaetos* in Finland. In *Status of Raptor Populations in Eastern Fennoscandia, Proceedings of the Workshop, Kostomuksha, Karelia, Russia, 8–10 November 2005*; Karelian Research Centre of the Russian Academy of Science Finnish-Russian Working Group on Nature Conservation: Petrozavodosk, Russia, 2005; pp. 114–116.
31. Ollila, T. *Raportti Maakotkan, Muuttohaukan, Tunturihaukan Sekä Oulun ja Lapin Läänien Merikotkien Pesinnöistä Vuonna 2019*; Metsähallitus: Vantaa, Finland, 2019. (In Finnish)
32. Ollila, T. *Raportti Maakotkan, Muuttohaukan, Tunturihaukan Sekä Oulun ja Lapin Läänien Merikotkien Pesinnöistä Vuonna 2020*; Metsähallitus: Vantaa, Finland, 2020. (In Finnish)
33. Ollila, T. Suomen maakotkat vuonna 2018—Onko pesintämenestys huonontumassa? *Linnut Vuosik.* **2019**, *2018*, 104–109, (In Finnish with English Summary).
34. Honkala, J.; Viikki, C. Europe's Oldest Golden Eagle is Finnish; Life Science News. University of Helsinki. 2019. Available online: https://www.helsinki.fi/en/news/life-science-news/europes-oldest-golden-eagle-is-finnish (accessed on 14 March 2023).
35. Tjernberg, M. Kungsörnens status och ekologi i Sverige 2006 samt tänkbara prognoser för artens urveckling. In *Rapport till Rovdjursutredningen—ArtDatabanken*; Swedish Agricultural University: Uppsala, Sweden, 2006. (In Swedish)
36. Brown, L.; Amadon, D. *Eagles, Hawks, and Falcons of the World*; Country Life Books: London, UK, 1968; Volume 2.
37. Slater, S.; Keller, K.; Knight, R. Interannual Golden Eagle *(Aquila chrysaetos)* Nest-Use Patterns in Central Utah: Implications for Long-Term Nest Protection. *J. Raptor Res.* **2017**, *51*, 129–135. [CrossRef]
38. Millsap, B.A.; Grubb, T.G.; Murphy, R.K.; Swem, T.; Watson, J.W. Conservation significance of alternative nests of Golden Eagles. *Glob. Ecol. Conserv.* **2015**, *3*, 234–241. [CrossRef]
39. Bielikova, M.; Ficek, A.; Valkova, D.; Turna, J. Multiplex PCR amplification of 13 microsatellite loci for *Aquila chrysaetos* in forensic applications. *Biologia* **2010**, *65*, 1081–1088. [CrossRef]
40. Nebel, C.; Gamauf, A.; Haring, E.; Segelbacher, G.; Villers, A.; Zachos, F.E. Mitochondrial DNA analysis reveals Holarctic homogeneity and a distinct Mediterranean lineage in the Golden Eagle (*Aquila chrysaetos*). *Biol. J. Linn. Soc.* **2015**, *116*, 328–340. [CrossRef]
41. Sonsthagen, S.A.; Coonan, T.J.; Latta, B.C.; Sage, G.K.; Talbot, S.L. Genetic diversity of a newly established population of Golden Eagles on the Channel Islands, California. *Biol. Conserv.* **2012**, *146*, 116–122. [CrossRef]
42. Griffiths, R.; Double, M.C.; Orr, K.; Dawson, R.J. A DNA test to sex most birds. *Mol. Ecol.* **1998**, *7*, 1071–1075. [CrossRef] [PubMed]
43. Ito, H.; Sudo-Yamaji, A.; Abe, M.; Murase, T.; Tsubota, T. Sex identification by alternative polymerase chain reaction methods in Falconiformes. *Zool. Sci.* **2003**, *20*, 339–344. [CrossRef] [PubMed]

44. Honka, J.; Merikanto, K. Microsat_Errcalc. 2020. Available online: https://github.com/karimerikanto/microsat_errcalc (accessed on 25 April 2022).
45. Kalinowski, S.T.; Taper, M.L.; Marshall, T.C. Revising how the computer program Cervus accommodates genotyping error increases success in paternity assignment. *Mol. Ecol.* **2007**, *16*, 1099–1106. [CrossRef]
46. van Oosterhout, C.; Hutchinson, B.; Wills, D.; Shipley, P. MICRO-CHECKER: Software for identifying and correcting genotyping errors in microsatellite data. *Mol. Ecol. Notes* **2004**, *4*, 535–538. [CrossRef]
47. Hall, T.A. BioEdit: A user-friendly biological sequence alignment editor and analysis program for Windows 95/98/NT. *Nucleic Acids Symp. Ser.* **1999**, *41*, 95–98.
48. Goudet, J.J. FSTAT (Version 1.2): A Computer Program to Calculate F-Statistics. *J. Hered.* **1995**, *86*, 485–486. [CrossRef]
49. Excoffier, L.; Lischer, H.E.L. Arlequin suite ver. 3.5: A new series of programs to perform population genetics analyses under Linux and Windows. *Mol. Ecol. Res.* **2010**, *10*, 564–567. [CrossRef]
50. Raymond, M.; Rousset, F. GENEPOP (version 1.2): Population genetics software for exact tests and ecumenicism. *Heredity* **1995**, *86*, 248–249. [CrossRef]
51. Pritchard, J.K.; Stephens, M.; Donnelly, P. Inference of population structure using multilocus genotype data. *Genetics* **2000**, *155*, 945–959. [CrossRef]
52. Li, Y.L.; Liu, J.X. StructureSelector: A web-based software to select and visualize the optimal number of clusters using multiple methods. *Mol. Ecol. Resour.* **2018**, *18*, 176–177. [CrossRef]
53. Evanno, G.; Regnaut, S.; Goudet, J. Detecting the Number of Clusters of Individuals Using the Software Structure: A Simulation Study. *Mol. Ecol.* **2005**, *14*, 2611–2620. [CrossRef] [PubMed]
54. Puechmaille, S.J. The Program Structure Does Not Reliably Recover the Correct Population Structure When Sampling Is Uneven: Subsampling and New Estimators Alleviate the Problem. *Mol. Ecol. Resour.* **2016**, *16*, 608–627. [CrossRef] [PubMed]
55. Francis, R.M. POPHELPER: An R package and web app to analyse and visualise population structure. *Mol. Ecol. Resour.* **2016**, *17*, 27–32. [CrossRef] [PubMed]
56. Thompson, J.D.; Higgins, D.G.; Gibson, T.J. CLUSTAL W: Improving the sensitivity of progressive multiple sequence alignment through sequence weighting, position-specific gap penalties and weight matrix choice. *Nucleic Acids Res.* **1994**, *22*, 4673–4680. [CrossRef]
57. Rozas, J.; Ferrer-Mata, A.; Sánchez-DelBarrio, J.C.; Guirao-Rico, S.; Librado, P.; Ramos-Onsins, S.E.; Sánchez-Gracia, A. DnaSP 6: DNA Sequence Polymorphism Analysis of Large Datasets. *Mol. Biol. Evol.* **2017**, *34*, 3299–3302. [CrossRef]
58. Bandelt, H.; Forster, P.; Röhl, A. Median-joining networks for inferring intraspecific phylogenies. *Mol. Biol. Evol.* **1999**, *16*, 37–48. [CrossRef]
59. Leigh, J.W.; Bryant, D. POPART: Full-feature software for haplotype network construction. *Methods Ecol. Evol.* **2015**, *6*, 1110–1116. [CrossRef]
60. Inkscape Project. Inkscape. Available online: https://inkscape.org (accessed on 25 February 2023).
61. RStudio Team. RStudio: Integrated Development for R. 2022. Available online: http://www.rstudio.com/ (accessed on 10 May 2022).
62. Paradis, E. Pegas: An R package for population genetics with an integrated–modular approach. *Bioinformatics* **2010**, *26*, 419–420. [CrossRef]
63. Paradis, E.; Schliep, K. Ape 5.0: An environment for modern phylogenetics and evolutionary analyses in R. *Bioinformatics* **2019**, *35*, 526–528. [CrossRef]
64. Kalinowski, S.T.; Wagner, A.P.; Taper, M.L. ML-Relate: A computer program for maximum likelihood estimation of relatedness and relationship. *Mol. Ecol. Notes* **2006**, *6*, 576–579. [CrossRef]
65. QGIS Development Team. QGIS Geographic Information System. Open Source Geospatial Foundation Project. 2023. Available online: http://qgis.osgeo.org (accessed on 14 March 2023).
66. Nebel, C.; Gamauf, A.; Haring, E.; Segelbacher, G.; Väli, Ü.; Villers, A.; Zachos, F.E. New insights into population structure of the European Golden Eagle (*Aquila chrysaetos*) revealed by microsatellite analysis. *Biol. J. Linn. Soc.* **2019**, *128*, 611–631. [CrossRef]
67. Craig, E.H.; Adams, J.R.; Waits, L.P.; Fuller, M.R.; Whittington, D.M. Nuclear and Mitochondrial DNA Analyses of Golden Eagles (*Aquila chrysaetos canadensis*) from Three Areas in Western North America; Initial Results and Conservation Implications. *PLoS ONE* **2016**, *11*, 164248. [CrossRef] [PubMed]
68. Bourke, B.P.; Frantz, A.C.; Lavers, C.P.; Davison, A.; Dawson, D.A.; Burke, T.A. Genetic Signatures of Population Change in the British Golden Eagle (*Aquila chrysaetos*). *Conserv. Genet.* **2010**, *11*, 1837–1846. [CrossRef]
69. Ogden, R.; Heap, E.; McEwing, R.; Tingay, R.; Whitfield, D.P. Population Structure and Dispersal Patterns in Scottish Golden Eagles *Aquila chrysaetos* Revealed by Molecular Genetic Analysis of Territorial Birds. *Ibis* **2015**, *157*, 837–848. [CrossRef]
70. Sato, Y.; Ogden, R.; Komatsu, M.; Maeda, T.; Inoue-Murayama, M. Integration of Wild and Captive Genetic Management Approaches to Support Conservation of the Endangered Japanese Golden Eagle. *Biol. Conserv.* **2017**, *213*, 175–184. [CrossRef]
71. Judkins, M.E.; van den Bussche, R.A. Holarctic Phylogeography of Golden Eagles (*Aquila chrysaetos*) and Evaluation of Alternative North American Management Approaches. *Biol. J. Linn. Soc.* **2018**, *123*, 471–482. [CrossRef]
72. Klug, H. Why monogamy? A review of potential ultimate drivers. *Front. Ecol. Evol.* **2018**, *6*, 30. [CrossRef]
73. Harvey, P.H.; Greenwood, P.J.; Campbell, B.; Stenning, M.J. Breeding Dispersal of the Pied Flycatcher (*Ficedula hypoleuca*). *J. Anim. Ecol.* **1984**, *53*, 727–736. [CrossRef]

74. Divoky, G.J.; Horton, M. Breeding and Natal Dispersal, Nest Habitat Loss and Implications for Marbled Murrelet Populations. In *Ecology and Conservation of the Marbled Murrelet*; USDA Forest Service General Technical Report PSW-GTR-152; Ralph, C.J., Hunt, G.L., Jr., Raphael, M.G.E., Piatt, J.F., Eds.; USDA Forest Service: Albany, CA, USA, 1995; pp. 83–88.
75. Suvantola, L. The Golden Eagle Compensation Scheme in Finland as an Example of Incentive Measures: Potential for Conflict Management? In *Human-Wildlife Conflicts in Europe: Fisheries and Fish-Eating Vertebrates as a Model Case*; Klenke, R.A., Ring, I., Kranz, A., Jepsen, N., Rauschmayer, F., Henle, K., Eds.; Springer: Berlin/Heidelberg, Germany, 2013; pp. 201–214.
76. Gowaty, P.A.; Karlin, A.A. Multiple maternity and paternity in single broods of apparently monogamous Eastern Bluebirds (*Sialia sialis*). *Behav. Ecol. Sociobiol.* **1984**, *15*, 91–95. [CrossRef]
77. Blakesley, J.A.; Anderson, D.R.; Noon, B.R. Breeding dispersal in the California Spotted Owl. *Condor* **2006**, *108*, 71–81. [CrossRef]
78. Blomqvist, D.; Andersson, M.; Küpper, C.; Cuthill, I.C.; Kis, J.; Lanctot, R.B.; Brett, K.; Sandercock, B.K.; Székely, T.; Wallander, J.; et al. Genetic similarity between mates and extra-pair parentage in three species of shorebirds. *Nature* **2002**, *419*, 613–615. [CrossRef] [PubMed]
79. Bao, X.; Liu, F.; Fan, Y.; Jia, B.; Li, J. Extra-pair paternity in two passerine species living in an extreme-arid desert region. *Wilson J. Ornithol.* **2019**, *131*, 638–644. [CrossRef] [PubMed]
80. Krams, I.A.; Mennerat, A.; Krama, T.; Krams, R.; Joers, P.; Elferts, D.; Luoto, E.; Rantala, M.J.; Eliassen, S. Extra-pair paternity explains cooperation in a bird species. *Proc. Natl. Acad. Sci. USA* **2022**, *119*, e2112004119. [CrossRef] [PubMed]
81. Laistal, O. *Im Tal der Königsadler: Mit den Zugvögeln zum Polarkreis*; Landbuck-Verklag GmbH: Hannover, Germany, 1966. (In German)
82. Dennis, R. Probable polygyny by Golden Eagle. *Br. Birds* **1983**, *76*, 310–311.
83. Bergo, G. Trios in the Golden Eagle *Aquila chrysaetos* (L.). *Fauna Norv. Ser. C Cindus* **1988**, *11*, 40–44.
84. Whitfield, D.P.; Fielding, A.H.; McLeod, D.R.A.; Haworth, P.F.; Watson, J. A conservation framework for the Golden Eagle in Scotland: Refining condition targets and assessment of constraint influences. *Biol. Conserv.* **2006**, *130*, 465–480. [CrossRef]
85. Culina, A.; Radersma, R.; Sheldon, B.C. Trading up: The fitness consequences of divorce in monogamous birds. *Biol. Rev.* **2015**, *90*, 1015–1034. [CrossRef]
86. Boeker, E.L.; Ray, T.D. Golden Eagle population studies in the Southwest. *Condor* **1971**, *73*, 463–467. [CrossRef]
87. Millon, A.; Arroyo, B.E.; Bretagnolle, V. Variable but predictable prey availability affects predator breeding success: Natural versus experimental evidence. *J. Zool.* **2008**, *275*, 349–358. [CrossRef]
88. Steenhof, K.; Kochert, M.N.; Mcdonald, T.L. Interactive effects of prey and weather on Golden Eagle reproduction. *J. Anim. Ecol.* **1997**, *66*, 350. [CrossRef]
89. Kaisanlahti-Jokimäki, M.L.; Jokimäki, J.; Huhta, E.; Ukkola, M.; Helle, P.; Ollila, T. Territory occupancy and breeding success of the Golden Eagle (*Aquila chrysaetos*) around tourist destinations in northern Finland. *Ornis Fenn.* **2008**, *85*, 2–12.
90. Watson, J.; Rae, S.R.; Stillman, R. Nesting density and breeding success of Golden Eagles in relation to food supply in Scotland. *J. Anim. Ecol.* **1992**, *61*, 543–550. [CrossRef]
91. Helle, P.; Ikonen, K.; Kantola, A. Wildlife monitoring in Finland: Online information for game administration, hunters, and the wider public. *Can. J. For. Res.* **2016**, *46*, 1491–1496. [CrossRef]
92. Martínez-Cruz, B.; David, V.A.; Godoy, J.A.; Negro, J.J.; O'Brien, S.J.; Johnson, W.E. Eighteen polymorphic microsatellite markers for the highly endangered Spanish imperial eagle (*Aquila adalberti*) and related species. *Mol. Ecol. Notes* **2002**, *2*, 323–326. [CrossRef]
93. Nesje, M.; Røed, K.H. Microsatellite DNA markers from the gyrfalcon (*Falco rusticolus*) and their use in other raptor species. *Mol. Ecol.* **2000**, *9*, 1438–1440. [CrossRef] [PubMed]

diversity

MDPI

Article

Substantial Variation in Prospecting Behaviour of Young Golden Eagles *Aquila chrysaetos* Defies Expectations from Potential Predictors

Alan H. Fielding [1], David Anderson [2], Stuart Benn [3], Robin Reid [4], Ruth Tingay [5], Ewan D. Weston [1] and D. Philip Whitfield [1,*]

1 Natural Research Ltd., Brathens, Aberdeenshire AB31 4BY, UK; a.h.fielding@gmail.com (A.H.F.); ewan_weston@hotmail.com (E.D.W.)
2 Dave Anderson Ecology Ltd., Callander FK17 8EU, UK; dikanderson@googlemail.com
3 RSPB Scotland, Inverness IV2 3BW, UK; stuart@polarfox.myzen.co.uk
4 Independent Researcher, Isle of Harris HS3 3EZ, UK; vikingrobin@hotmail.com
5 Wild Justice, 9 Lawson Street, Raunds, Northants NN9 6NG, UK
* Correspondence: phil.whitfield@natural-research.org

Citation: Fielding, A.H.; Anderson, D.; Benn, S.; Reid, R.; Tingay, R.; Weston, E.D.; Whitfield, D.P. Substantial Variation in Prospecting Behaviour of Young Golden Eagles *Aquila chrysaetos* Defies Expectations from Potential Predictors. *Diversity* **2023**, *15*, 506. https://doi.org/10.3390/d15040506

Academic Editors: Michael Wink, Vladimir D. Dobrev, Dobromir D. Dobrev and Dimitar A. Demerdzhiev

Received: 21 February 2023
Revised: 24 March 2023
Accepted: 27 March 2023
Published: 1 April 2023

Abstract: Natal dispersal of large raptors is poorly understood, despite being a crucial transitional life-cycle phase affecting gene flow and population dynamics. A research-gap rarely examined concerns how young dispersing raptors strategize movements towards first settlement on a prospective breeding territory. First territory settlement is a critical decision for a lifetime and can take several years. With such importance, large raptors should theoretically devote considerable effort to acquire accurate information during prospection. Nevertheless, when discovery of territorial opportunities may be difficult, but easier in vacant territories, we posit two extremes in strategizing prospection behaviours: (1) "Quick, grab it when available with limited reconnaissance", as opposed to (2) "Slow, waiting game with frequent reconnaissance". We analysed pre-settlement data from 37 GPS-tagged nestling golden eagles, later recorded as having settled on their first territory. The number of eagles' unique daily visits to their later settled territories was a measure of prospection intensity. We documented substantial variation in prior visits, between less than 10 to several hundred. Analyses considered several potential predictors. We expected a positive association between number of prospecting visits and natal dispersal duration, since with more time to gather information there should be more visits. We also expected fewer prospecting visits in prior vacant territories. Neither of these expectations were supported. There was a non-significant tendency for more prospection visits by males. Our study provides novel information on a seldom-studied behaviour in a large raptor. It illustrates substantial variation in prospecting behaviour, but expectations of potential drivers behind this variation were not confirmed, urging further study.

Keywords: reconnaissance behaviour; natal dispersal; juvenile dispersal; transience phase; raptor; population ecology; GPS-telemetry

1. Introduction

Much of the behaviour of large raptors during natal dispersal [1,2], often termed juvenile dispersal [3–7], is poorly known [8]. This is the phase between birds departing from their natal range (at the end of post-fledging dependence phase: PFDP [7,9]) and settlement on a prospective breeding territory [10]. A particular research-gap in this critical phase, otherwise termed as transience [7,11–14], involves how dispersing birds prospect or may strategize their first settlement on a breeding territory [14,15].

First territory settlement in large raptors is important because it is substantially a decision for a lifetime via territory quality, available partner, and hence reproductive resources. After initial settlement, further movements (via breeding dispersal: [1]) appear uncommon in most species [10,16,17]. While large raptors can take several years during natal dispersal to decide on or find a prospective breeding territory in which to settle, natal dispersal is variable in duration [10].

Young large raptors face challenges in prospecting and finding any available territorial opportunity which may lead to reproduction. Life history traits, notably high longevity, and low density as top predators [16], can produce a conspecific landscape into which youngsters try to establish themselves as a reproductive entity that is typically thinly spread, usually filled with defensive occupants, and with few territorial openings for easy entry [10]. Those openings are also sought by other prospecting youngsters. Existing territorial birds, by regularly visiting neighbouring territories, likely scoping an advantageous movement, are an additional competitive element faced by dispersing youngsters [10,18]. Such opportunities may be so low in high density populations that older adults may still be 'floaters' seeking settlement where reproduction is possible [19–21].

From research on other species, prospecting future breeding opportunities may be arguably easier in species which are cavity nesting or clumped in distribution, such as colonial seabirds [15,22–25], even though prospection studies often involve breeding dispersal [26,27].

There are few studies of large raptors. Prospecting behaviour in young eagle owls *Bubo bubo* was uncommon [28]. Dispersing owls occupied long-term 'temporary settlement areas' (TSAs) close to breeding ranges with minimal prospecting, even though a few young owls rapidly occupied a breeding territory [28]. While revelatory, these features may be unusual. In other large raptors, such as the golden eagle *Aquila chrysaetos*, superficially there are far greater movements during natal dispersal which involve potential prospection, with shorter occupancy of TSAs [5,6,10,14,29].

Information gathering during prospection can be costly in time and energy [8,22], and may include death according to golden eagle studies [29–31]. Theory expects that, given the settlement decision's importance, birds should devote considerable effort to acquiring accurate information about potential locations before finally selecting a breeding site [23]. At some point, however, potential breeders must gain information regarding prospective breeding sites without the costs of gathering the information outweighing the benefits [32].

Scottish golden eagles settled when younger in prior vacant territories than in prior occupied territories, so time taken to settle was a function of the competitive environment [10]. This environment may also be relevant to prospecting behaviours under a cost-benefit evaluation [32]. Birds may nevertheless be constrained spatially in their information gathering movements by an innate natal dispersal distance (NDD: [1]) exerting a philopatric pull towards natal sites, potentially affecting prospecting movement behaviours. This philopatric pull is expected to be greater in males than females [1,2,33,34] and was confirmed in our study population [35].

Within this NDD constraint, and with a Scottish territorial landscape which can vary geographically between several vacant openings but with a higher density of occupied territories elsewhere [10,29], two respective extremes in strategizing prospection of territory availability may be: (1) "Quick, grab it when available with limited reconnaissance", as opposed to (2) "Slow, waiting game with frequent reconnaissance".

In simple cost-benefit analysis, tactic 1 avoids much prospecting cost but may not benefit subsequently from reconnoitering the best opportunity available within a limited time span. Tactic 2 encompasses the opposite strategy. These two extremes may be relevant to prospecting behaviours because in our study system eagles settled on a prospective breeding territory when younger if it was vacant. They took longer to settle on a territory which was occupied [10]. These results came from satellite telemetry research. Satellite telemetry provides novel research opportunities in avian ecology and behaviour [36], including tracking prospecting movements [15], although in large raptors it has been seldom utilised [14].

Using records of golden eagles GPS-tagged as nestlings which later settled on a territory [10], our objectives were first to describe the number of unique daily visits made during natal dispersal to the later settled territory, as a measure of prospection intensity. Under a second objective, this variable was set analytically against several potential predictors which we expected could be influential in affecting our prospection metric.

2. Methods

2.1. Study Area and Species

Scotland covers c. 80,000 km^2 on the northwestern limit of Europe and hosts around 500 territorial golden eagle pairs occupying Scotland's uplands [37], which are also used by non-territorial birds [29,38–40]. These uplands vary in geology, vegetation, topography and climatic influences.

Climatically, situated on the northeastern edge of the Atlantic Ocean, the west of Scotland being subjected more to the Atlantic Drift is wetter and windier with more equitable seasonal changes in weather, and is hence more oceanic. The east is drier with greater seasonal change in weather and is more continental [38,41]. The contrasting oceanic/continental influences tend to produce upland vegetations which are found only at higher altitude in the east but which may occur at sea level in the west. The preferred open habitats are vegetated by dry or wet heathland and peatland dominated by heather *Calluna vulgaris* and relatives in the east, with graminoids, sedges and deeper peatland more common in the west [38,41].

There is a broad west/east divide on the strong influence of illegal persecution of golden eagles. In the east, driven shoots of red grouse (*Lagopus lagopus scotica*) are practiced and require management for extremely high densities of grouse coveys [38,42–46]. To maintain such high densities, some land managers illegally kill predators of grouse such as golden eagles. Illegal persecution of golden eagles has been evident for several decades. These illegal activities substantially contribute to a national environment of numerous vacant eagle territories and a relatively low density in the east, compared to higher densities further west [29,37,38,41–46]. Such eastern areas can be attractive to eagles, but as they can be lethal, this provides a classic example of an "ecological black hole" due to anthropogenic activities [38,43,44].

The west/east influences are also reflected in golden eagles' diet in being more restricted to red grouse and mountain hares *Lepus timidus* in the east [41,47]. Despite large numbers of hares being culled on some intensively managed grouse moors [48] the abundance of red grouse and hares in the east is often associated with higher breeding productivity there, from the fewer territories which prevail—even if only sometimes temporarily [38,41–47]. In the oceanic west, climatic influences may explain much productivity variation [49], although a low dietary diversity ('specialization') was not associated with higher productivity [50].

Golden eagles in Scotland do not use lowland habitats [39,40]. They do use low altitude habitats in the western Highlands and Islands, but given the strong oceanic influences here, the "uplands" can often descend to sea-level. Topographic measures, as a surrogate for availability of orographic/declivity, winds and aerial habitat facilitating movement involving combined measures of altitude, slope and distance from ridge, are powerful predictors of golden eagle activity [39].

2.2. GPS Satellite Tagging

Tagging methods for Scottish golden eagles have been repeatedly described [7,9,29,39,40,51]. Nestlings were tagged when 50–70 days old [52,53] with transmitter weights and harnesses less than the 3% lower recommended maximum of body weight [54] (see also [55,56]). Using 13 mm tubular Teflon ribbon (Bally Ribbon Mills, Bally, PA, USA) sewed with cotton or linen thread, all birds were fitted with a thoracic X strap harness [57,58], otherwise called a "crossover wing harness" [59] or a "Garcelon-type harness" [60]. This method, and others associated with tagging, were followed [58,61]. This harness design did not affect several raptors on survival or physical injuries, including golden eagle [60]. Tagging of Scottish eagles had no apparent adverse effects under these metrics and others, including breeding productivity [29].

Several solar-charged transmitter models were deployed [9,29]. Almost all deployed tags in the present study were manufactured by MTI (Microwave Telemetry Inc., Columbia, MD, USA), and their specifications and transmission outputs are described elsewhere [7,9,10,29,39,40,51].

2.3. Estimating Variables and Statistical Analyses

Territory settlement was determined algorithmically, involving spatial and temporal evaluations accounting for potentially confounding TSA use and exploratory movements by settled territorial birds [10]. The same telemetry data were used to estimate the territory's 95% percent volume contour (PVC) using the kernel density estimator in the adehabitatHR package in R [62]. We refer to this subsequently as the range.

We recorded the number of unique days when the range was visited prior to settlement (after the post-fledging dependence period: PFDP [7,9]) using available daily points ascertained by the polygon tool of QGIS 3.26. This was taken as our measure of prospection intensity, in description and as the dependent variable in analyses. We did not use the number of records in the range as this metric would be more affected by the state of a tag's battery and its solar recharge, which can decline in winter and in darkness [63]. Nocturnal records were excluded using the R suncalc package (v 0.5.0) [64].

We utilized five potential predictors (explanatory variables) in our analyses:

1. The duration of natal dispersal taken as dates between when an eagle dispersed (after PFDP: [7]) and when it settled on a territory [10]. This was expected to be influential, since with longer natal dispersal, opportunities for more prospection visits were greater.
2. The prior status of the settled territory: occupied or vacant [10]. Under our premise on 'quick' or 'slow' reconnaissance (Introduction), the ease of prospection and settlement should be greater in vacant territories—towards the 'quick' premise.
3. The duration to first date after dispersing (i.e., post-PFDP) when an eagle first visited its later settled territory. We included this variable under the premise that early experience of the settled territory may have had an influence on how quickly later settlement occurred.
4. The duration of PFDP. The beginning of PFDP for birds tagged as nestlings was taken as a generic fledging date of 1 August, to first date of dispersal on duration [7,9]. These measures gave the start and end dates of the metric. This metric was included as a potential predictor because PFDP is highly variable [7,9] and if birds spend more time in their natal territory, then this could have consequences for their later prospection efforts.
5. Sex. Differences between sexes surrounding natal dispersal are several [1,2,7,9,10,33–35].

Using R (v 4.2.3), we computed single predictor general linear models (GLMs) with Gaussian error distribution. Predictor significance was obtained using the R summary command. Given that there were five potential predictors, there were many potential multi-predictor models. We used the dredge function from the R MuMIn package (v 1.47.1) for multi-model inference and used AIC to rank the models. We selected the model with the lowest AIC as the best model but we acknowledged a warning [65] about selecting a spurious best model, and therefore we included parsimony as an important criterion.

3. Results

Our study involved 37 tagged eagles (Table 1). Summary descriptive statistics showed that there was a wide variation in the number of days when young eagles visited their future territorial range prior to settlement, from one to 493 (minimum and maximum values) (Table 2).

We fitted GLM models, estimated using maximum likelihood, using prior unique daily visits to the future range as the dependent and the five potential predictors (Methods) as prospective explanatories. None of the predictors was significant when used in isolation (Table 3). We also fitted a full model but its explanatory power to predict the number of visits to the future range was weak ($R^2 = 0.09$). The full model's intercept, corresponding to sex = F, prior status = occupied, days to first visit = 0, natal dispersal duration = 0 and PFDP duration = 0, was at 118.41 (95% CI [−71.27, 308.09], t (df = 31) = 1.22, $p = 0.221$).

Table 1. Raw data used in analyses. Tag = Tag ID, Sex (M = male, F = female), Dispersal = date of dispersal (end of PFDP: d/m/y), PFDP = duration of PFDP (d), 1st visit = date of first visit to future range, Day 1 = 1st visit to future range (d) where 0 = start of dispersal, Visits = number of unique dates (d) on which the settled range was visited, Settled = date the bird settled (d/m/y), Duration = natal dispersal duration (d) (number of days to settlement from the start of dispersal, day 0), NDD = Distance from natal site to future range (km), Prior Status = status of the future range prior to settlement (O = occupied, V = vacant).

Tag	Sex	Dispersal	PFDP	1st Visit	Day 1	Visits	Settled	Duration	NDD	Prior Status
100	M	04/02/15	173	14/03/15	38	187	22/03/19	1507	60	O
102	F	01/11/08	78	09/11/08	8	7	07/02/12	1193	58.6	O
334	M	09/12/16	116	12/04/17	124	64	25/09/18	655	26.9	V
582	F	07/11/16	84	10/04/17	154	67	06/03/21	1580	60	O
584	M	12/04/16	241	13/04/17	366	283	16/10/21	2013	18	V
660	M	01/03/18	198	18/04/18	48	70	06/01/19	311	13.6	V
809	F	22/09/16	38	04/04/17	194	32	24/05/20	1340	63	O
815	F	04/03/09	201	05/10/11	945	1	01/03/14	1823	25	O
928	M	03/02/18	172	25/02/18	22	31	29/04/21	1181	36	O
932	M	21/03/18	218	02/08/19	499	48	07/10/19	565	43.1	V
933	F	19/09/19	32	14/09/20	361	9	20/04/22	944	50.8	V
1025	M	20/01/19	158	14/03/19	83	40	10/09/21	964	35	V
1026	M	11/03/19	208	20/01/20	315	38	08/12/21	1003	52	V
1030	F	28/10/18	74	09/04/19	163	73	19/04/21	904	68	V
1031	M	01/04/19	229	02/04/19	1	461	06/09/20	524	66	V
1074	F	31/10/19	77	02/03/20	123	315	07/02/21	465	83	V
1094	M	06/12/19	113	26/04/20	142	226	17/12/21	742	81.9	V
1096	F	30/10/19	76	21/11/19	22	402	19/05/21	567	12	O
1097	M	30/10/19	76	25/04/22	908	493	29/08/22	1034	33.1	O
21197	F	28/12/10	135	02/05/11	125	7	23/01/13	757	78	O
51888	M	03/03/15	200	05/09/15	186	31	04/02/18	1069	27	O
57109	M	28/09/10	44	08/02/11	133	357	20/02/15	1606	33	O
57115	M	17/10/10	63	27/03/11	161	42	23/10/11	371	30	V
84135	F	17/10/10	63	28/06/11	254	54	15/12/11	424	82	V
89251	F	05/02/12	174	27/06/13	508	13	17/12/14	1046	65	O
89279	F	17/12/11	124	11/02/12	56	114	16/02/16	1522	38	O
120196	M	04/11/12	81	26/03/14	507	8	10/04/15	887	29	O
129005	M	06/11/13	83	08/11/13	2	20	21/01/16	806	26	V
129006	M	05/10/13	51	26/10/13	21	110	15/01/19	1928	26	O
129008	F	31/10/14	77	02/06/15	214	26	12/11/16	743	32	V
129012	M	05/12/13	112	11/01/14	37	301	07/04/15	488	13	V
148632	F	28/11/15	105	27/03/16	120	64	14/02/17	444	40	V
148635	F	25/10/15	71	04/03/17	496	2	10/03/17	502	87	V
148639	F	11/02/16	180	20/04/16	69	65	02/05/17	446	56	V
148640	M	05/02/16	174	21/02/16	16	100	20/02/17	381	42	V
286611	F	10/11/07	87	13/03/09	489	2	05/04/09	512	67	V
656352	M	22/03/09	219	14/04/09	23	65	31/03/12	1105	29	O

Table 2. Descriptive statistics for the explanatory variable (home range visits: number of unique days), and predictive variables: natal dispersal duration, first visit to the future range and duration of PFDP. These are cast according to sex (M = male F = female) and the status of the future range prior to settlement (occupied or vacant). LCL = 2.5% quantile, UCL = 97.5% quantile.

Variable	Sex	Prior Status	Min	LCL	Mean	Median	UCL	Max	sd
Home range visits (d)	F	occupied	1	2	80	23	352	402	136
	F	vacant	2	2	68	54	267	315	97
	M	occupied	8	12	160	88	469	493	177
	M	vacant	20	25	141	67	417	461	142
Natal dispersal duration (d)	F	occupied	567	600	1228	1266	1780	1823	426
	F	vacant	424	428	598	502	936	944	208
	M	occupied	887	913	1290	1143	1872	1928	354
	M	vacant	311	328	735	610	1735	2013	461
First visit to future range (d)	F	occupied	8	11	252	140	869	945	322
	F	vacant	69	79	254	214	495	496	160
	M	occupied	21	21	230	86	838	908	319
	M	vacant	1	1	150	104	462	499	161
Duration of PFDP (d)	F	occupied	38	45	114	104	196	201	55
	F	vacant	32	38	85	77	165	180	41
	M	occupied	44	45	127	126	216	219	71
	M	vacant	63	69	159	166	238	241	61

Table 3. Results of individual Gaussian GLMs estimated by maximum likelihood using number of visits to the future range as the response (dependent) variable. Prospective explanatory predictor variables involved sex (male or female), duration of natal dispersal, prior occupancy status (occupied or vacant), days to first visit after dispersing and duration of PFDP.

Predictor	Beta	95% CI: Lower, Upper	*t* (df)	Standard Beta (95% CI)	*p*
Sex (corresponding to F)					
Sex (M)	75.04	−12.84, 162.93	1.67 (35)	0.54 (−0.09, 1.17)	0.094
Natal dispersal duration (corresponding to zero)					
Natal dispersal duration (d)	0.0	0.08, 0.11	0.29 (35)	0.05 (−0.28, 0.38)	0.775
Prior occupancy status (corresponding to vacant)					
Prior occupancy status (occupied)	−10.65	−102.45, 81.16	−0.23 (35)	−0.08 (−0.74, 0.58)	0.820
First day to range visit (corresponding to zero)					
First visit to future range (d)	0.00	−0.09, 0.11	0.16 (35)	0.03 (−0.30, 0.36)	0.870
Duration of PFDP (corresponding to zero)					
Duration of PFDP (d)	−0.73	−0.73, 0.75	0.03 (35)	0.01 (−0.33, 0.34)	0.976

There were no significant results indicating any potential predictor variable's influence (Table 3). An influence of sex was suggested (Table 3: $p = 0.094$) such that males tended to make more prospection visits to the future range and had a greater spread of prior visitation efforts than females (Figure 1).

Multi-model inference using the dredge function identified the highest ranked model, out of 16, as one with sex as the only predictor approaching possible influence. Sex was insignificant in this single predictor model, however, and the adjusted R^2 (0.032) was weak. This highest ranked model had an AIC of 535.8, while the second highest ranked model was the null model with an AIC of 536.2 (delta AIC = 0.42). Using parsimony, and the absence of a significant predictor in the highest ranked model, the null model was selected as the 'best' model. Hence, the absence of significant predictors, out of those tested, was confirmed.

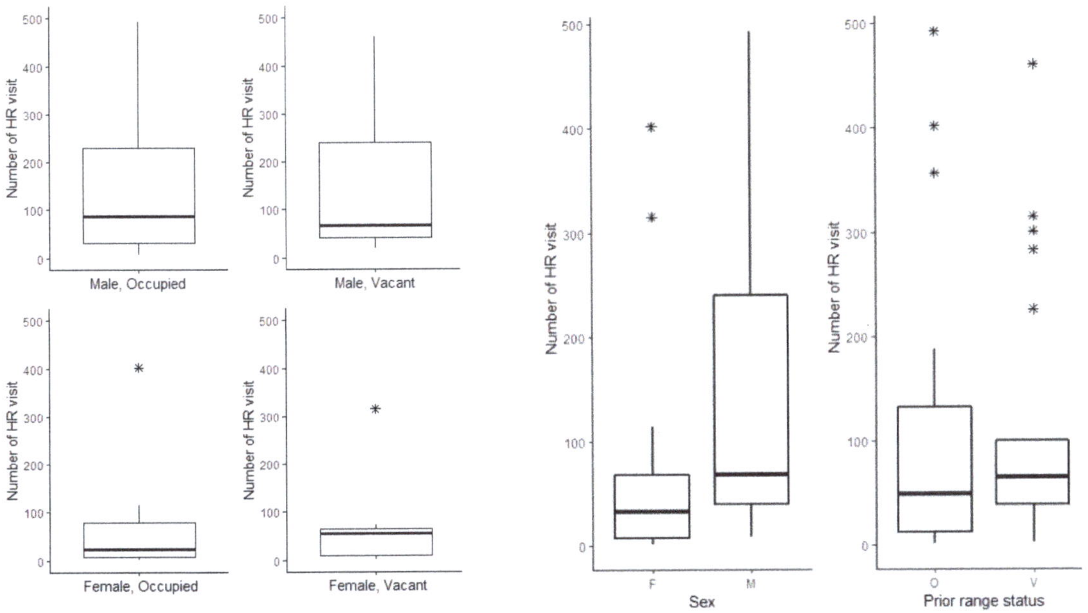

Figure 1. Illustrations of some results from analyses, with number of daily prior visits to the home range as the dependent variable (y axes). Presentations are box and whisker plots according to selected independent predictor variables (x axes). Solid black horizontal lines show medians, and box limits correspond to the first and third quartiles (25 and 75 percentiles). Whiskers (vertical lines) show values extending 1.5 times away from this inter-quartile range, and asterisks show values which were greater. The upper panels show results for sex and prior range status. With the suggestion of some influence of sex (upper left graphic), the middle panels show results for males according to prior range status (vacant or occupied), with corresponding results for females in the bottom panels.

4. Discussion

Golden eagles showed a wide range of prospection behaviours in how often the future settled territory was visited during natal dispersal, from a handful of visits to hundreds. This is a novel result in a field seldom considered previously in large raptors [14]. It is different to eagle owls in southern Spain [28].

We did not find any relationships with the large variation in our prospection metric. Two reasonable expectations were not supported as influential. Duration of natal dispersal (allowing more time for prospection) was not related. Prior occupancy status (if vacant then ease of settlement should produce fewer visits) was also not related, even though eagles settle when younger in vacant territories [10]. Such youthful settlement would expect fewer prospection visits but was not supported.

We posited two extremities for eagles' strategizing prospection behaviours: (1) "Quick, grab it when available with limited reconnaissance", as opposed to (2) "Slow, waiting game with frequent reconnaissance". Broadly, this was recorded. Tactic 1 should be more likely in vacant territories, however, and tactic 2 should be more likely in occupied territories. In some environments, the best territories should be more in demand and thus subject to more competition with a potentially longer waiting list for occupancy of higher quality territories [66]. This would result in more time to discover a territorial opening and more visits to reconnoiter any opening in higher quality territories.

This was not our study's environment, however. Critical to our study is that it involved a territorial landscape which was heavily influenced by illegal persecution, which has created many territorial vacancies in parts of eastern regions (Methods). We did not have data to explore settlement and prior prospecting behavioural decisions against subsequent reproductive outputs from settled birds, not least as age of occupants can affect reproductive success [43], and age of settlement is also variable and often young [10]. Nonetheless, in eastern regions where fewer territories are occupied, reproductive outputs are typically higher [38,45,46]. The waiting list [66] for such high-quality territories is therefore disproportionately shortened because of illegal persecution.

Hence, eagles settle when younger on these territories [10], and counter-intuitively, through effects of persecution, higher quality (vacant) territories were consequently expected to have a shorter waiting list with less reconnaissance during prospection. This was not found, however.

We also found no evidence that the rapidity with which an eagle first visited its later territory had any relationship with how often it was visited subsequently. Nor was there any relationship between the duration of PFDP—which can be highly variable [7,9]—with the substantial variation in prospection visits to the settled territory.

There was a non-significant indication that males made more prospection visits than females. This could be explained if males were more pioneering and active in establishing a new territory, since there were numerous vacant territories in our study system ([10,29,35], present study: Table 1). The role of females would arguably be more attuned to selecting a territory and their prospective male partner, and so with slightly later settlement. In a re-introduced population of white-tailed eagles *Haliaeetus albicilla*, this sequence of events was indicated [17]; however, see [67]. An earlier analysis of golden eagles in Scotland, however, did not find earlier settlement of males, even in vacant territories [10]. Including the present study, nonetheless, there is increasing evidence of differences between males and females, at stages involving natal dispersal which are still to be explained [9,10,35].

It is inevitable that dispersing birds will have visited several occupied ranges, and it is possible that the location and number of occupied ranges visited played a role in an individual's decision to settle. However, we were unable to investigate this because we did not have contemporaneous complete records of all occupied ranges across Scotland. This is a weakness in our study which is unlikely to be overcome in any reasonably sized population of large raptors.

In conclusion, we showed that young golden eagles had marked extremes in their reconnaissance tactics, completing very few to hundreds of prospection visits. We did not discover any significant predictors behind this variation, despite reasonable expectations under several potential explanatory variables. This illustrates that for large raptors it is a research field deserving more attention, if only initially on simple descriptive estimates of prospection behaviours [14].

Author Contributions: Conceptualization, A.H.F. and D.P.W.; methodology, all authors; validation, A.H.F. and D.P.W.; formal analysis, A.H.F. and D.P.W.; investigation, all authors; data curation, A.H.F.; writing-original draft preparation, A.H.F. and D.P.W.; writing-review and editing, all authors; visualization, all authors; supervision, A.H.F. and D.P.W.; project administration, D.P.W.; funding acquisition, D.P.W., R.T., D.A. and S.B. All authors have read and agreed to the published version of the manuscript.

Funding: Funding of tags and data download costs notably came from Natural Research, the Royal Society for the Protection of Birds, Roy Dennis Wildlife Foundation, Ruth Tingay and SSE Renewables. We are grateful to Emma Ahart, Thomas Plant, Nicki Small and Jenny Chambers (SSE Renewables) and Peter Robson (Scottish Power Renewables) for gaining permission to use data and to encourage tag funding, and to Chris Donald for a few tags funded by SNH. Two tags were provided by Movetech. Manuscript production was financially supported by SSE under the research programme of the Regional Eagle Conservation Management Plan. For facilitating this continued support, we thank Nicki Small and Jenny Chambers most recently. Despite this support, SSE had no influence or commentary in the production of the manuscript. Funding for open access publication was provided by Natural Research Ltd.

Institutional Review Board Statement: Not applicable.

Informed Consent Statement: Not applicable.

Data Availability Statement: Raw data are presented in Table 1. We cannot publicly or responsibly provide grid references for natal nest sites or subsequent settled territorial nest sites through prohibition under UK legislative practice.

Acknowledgments: Tagging was undertaken by David Anderson, Roy Dennis, Brian Etheridge, Justin Grant, Duncan Orr-Ewing and Ewan Weston: all were appropriately licensed under disturbance, handling, ringing and tagging licences from SNH (Scottish Natural Heritage: latterly NatureScot) and British Trust for Ornithology (BTO). We are extremely grateful for the considerable supporting fieldwork from many members of the Scottish Raptor Study Group. Staff at MTI (tag manufacture and support) and BTO (licensing) were helpful.

Conflicts of Interest: The authors declare no conflict of interest.

References

1. Greenwood, P.J. Mating systems, philopatry and dispersal in birds and mammals. *Anim. Behav.* **1980**, *28*, 1140–1162. [CrossRef]
2. Greenwood, P.J.; Harvey, P.H. The natal and breeding dispersal of birds. *Annu. Rev. Ecol. Syst.* **1982**, *13*, 83–88. [CrossRef]
3. Ferrer, M. Ontogeny of dispersal distances in young Spanish imperial eagles. *Behav. Ecol. Sociobiol.* **1993**, *32*, 259–263. [CrossRef]
4. Ferrer, M. Juvenile dispersal behavior and natal philopatry of a long-lived raptor, the Spanish Imperial Eagle *Aquila adalberti*. *Ibis* **1993**, *135*, 132–138. [CrossRef]
5. Whitfield, D.P.; Duffy, K.; McLeod, D.R.A.; Evans, R.J.; MacLennan, A.M.; Reid, R.; Sexton, D.; Wilson, J.D.; Douse, A. Juvenile dispersal of White-tailed Eagles in western Scotland. *J. Raptor Res.* **2009**, *43*, 110–120. [CrossRef]
6. Weston, E.D. Juvenile Dispersal Behaviour in the Golden Eagle (*Aquila chrysaetos*). Ph.D. Thesis, University of Aberdeen, Aberdeen, UK, 2014.
7. Weston, E.D.; Whitfield, D.P.; Travis, J.M.T.; Lambin, X. When do young birds disperse? Tests from studies of golden eagles in Scotland. *BMC Ecol.* **2013**, *13*, 42. [CrossRef]
8. Clobert, J.; Le Galliard, J.F.; Cote, J.; Meylan, S.; Massot, M. Informed dispersal, heterogeneity in animal dispersal syndromes and the dynamics of spatially structured populations. *Ecol. Lett.* **2009**, *12*, 197–209. [CrossRef]
9. Weston, E.D.; Whitfield, D.P.; Travis, J.M.T.; Lambin, X. The contribution of flight capability to the post-fledging dependence period of golden eagles. *J. Avian Biol.* **2018**, *49*, e01265. [CrossRef]
10. Whitfield, D.P.; Fielding, A.H.; Anderson, D.; Benn, S.; Dennis, R.; Grant, J.; Weston, E.D. Age of first territory settlement of Golden Eagles *Aquila chrysaetos* in a variable competitive landscape. *Front. Ecol. Evol.* **2022**, *10*, 743598. [CrossRef]
11. Stenseth, N.C.; Lidicker, W.Z. The study of dispersal: A conceptual guide. In *Animal Dispersal: Small Mammals as a Model*; Stenseth, N.C., Lidicker, W.Z., Jr., Eds.; Chapman and Hall: London, UK, 1992; pp. 5–20.
12. Delgado, M.M.; Penteriani, V. Behavioral states help translate dispersal movements into spatial distribution patterns of floaters. *Am. Nat.* **2008**, *172*, 475–485. [CrossRef]
13. Delgado, M.M.; Penteriani, V.; Nams, V.O.; Campioni, L. Changes of movement patterns from early dispersal to settlement. *Behav. Ecol. Sociobiol.* **2009**, *64*, 35–43. [CrossRef]
14. Poessel, S.A.; Woodbridge, B.; Smith, B.W.; Murphy, R.K.; Bedrosian, B.E.; Bell, D.A.; Bittner, D.; Bloom, P.H.; Crandall, R.H.; Domenech, R.; et al. Interpreting long-distance movements of non-migratory golden eagles: Prospecting and nomadism? *Ecosphere* **2022**, *13*, e4072. [CrossRef]
15. Ponchon, A.; Grémillet, D.; Doligez, B.; Chambert, T.; Tveraa, T.; González-Solís, J.; Boulinier, T. Tracking prospecting movements involved in breeding habitat selection: Insights, pitfalls and perspectives. *Methods Ecol. Evol.* **2013**, *4*, 143–150. [CrossRef]
16. Newton, I. *Population Ecology of Raptors*; Poyser: Berkhamstead, UK, 1979.
17. Whitfield, D.P.; Douse, A.; Evans, R.J.; Grant, J.; Love, J.; McLeod, D.R.A.; Reid, R.; Wilson, J.D. Natal and breeding dispersal in a reintroduced population of White-tailed Eagles *Haliaeetus albicilla*. *Bird Study* **2009**, *56*, 177–186. [CrossRef]
18. Watson, J.E.; Duff, A.A.; Davies, R.W. Home range and resource selection by GPS-monitored adult Golden Eagles in the Columbia plateau ecoregion: Implications for wind power development. *J. Wildl. Manag.* **2014**, *78*, 1012–1021. [CrossRef]
19. Hunt, W.G. Raptor floaters at Moffat's equilibrium. *Oikos* **1998**, *82*, 191–197. [CrossRef]
20. Penteriani, V.; Otalora, F.; Ferrer, M. Floater survival affects population persistence. The role of prey availability and environmental stochasticity. *Oikos* **2005**, *108*, 523–534. [CrossRef]
21. Sergio, F.; Tavecchia, G.; Blas, J.; López, L.; Tanferna, A.; Hiraldo, F. Variation in age-structured vital rates of a long-lived raptor: Implications for population growth. *Basic Appl. Ecol.* **2011**, *12*, 107–115. [CrossRef]
22. Reed, J.M.; Oring, L.W. Reconnaissance for future breeding sites by Spotted Sandpipers. *Behav. Ecol.* **1992**, *3*, 310–317. [CrossRef]
23. Reed, J.M.; Boulinier, T.; Danchin, E.; Oring, L.W. Informed Dispersal. In *Current Ornithology*; Nolan, V., Ketterson, E.D., Thompson, C.F., Eds.; Springer: Boston, MA, USA, 1999; Volume 19, pp. 189–259. [CrossRef]
24. Schjørring, S.; Gregersen, J.; Bregnballe, T. Prospecting enhances breeding success of first-time breeders in the great cormorant, *Phalacrocorax carbo sinensis*. *Anim. Behav.* **1999**, *57*, 647–654. [CrossRef]
25. Pärt, T.; Arlt, D.; Doligez, B.; Low, M.; Qvarnström, A. Prospectors combine social and environmental information to improve habitat selection and breeding success in the following year. *J. Anim. Ecol.* **2011**, *80*, 1227–1235. [CrossRef] [PubMed]

26. Ponchon, A.; Chambert, T.; Lobato, E.; Tveraa, T.; Grémillet, D.; Boulinier, T. Breeding failure induces large scale prospecting movements in the black-legged kittiwake. *J. Exp. Mar. Biol. Ecol.* **2015**, *473*, 136–145. [CrossRef]
27. Kloskowski, J. Win-stay/lose-switch, prospecting-based settlement strategy may not be adaptive under rapid environmental change. *Sci. Rep.* **2021**, *11*, 570. [CrossRef] [PubMed]
28. Fasciolo, A.; Delgado, M.M.; Cortés, G.; Soutullo, A.; Penteriani, V. Limited prospecting behaviour of juvenile Eagle Owls *Bubo bubo* during natal dispersal: Implications for conservation. *Bird Study* **2016**, *63*, 128–135. [CrossRef]
29. Whitfield, D.P.; Fielding, A.H. *Analyses of the Fates of Satellite Tracked Golden Eagles in Scotland*; SNH Commissioned Report No. 982; SNH: Battleby, UK, 2017.
30. Haller, H. Raumorganisation und dynamik einer population des Steinadlers *Aquila chrysaetos* in den zentralalpen. *Ornithol. Beob.* **1982**, *79*, 163–211.
31. Haller, H. Der Steinadler *Aquila chrysaetos* als Brutvögel im schweizerischen Alpenvorland: Ausbreitungstendenzen und ihre populationsökologischen Grundlagen. *Ornithol. Beob.* **1994**, *91*, 237–254.
32. Johnson, M.L. Exploratory behavior and dispersal: A graphical model. *Can. J. Zool.* **1989**, *67*, 2325–2328. [CrossRef]
33. Millsap, B.A.; Harmata, A.R.; Stahlecker, D.W.; Mikesic, D.G. Natal dispersal distance of bald and golden eagles originating in the coterminous United States as inferred from band encounters. *J. Raptor Res.* **2014**, *48*, 13–23. [CrossRef]
34. Murphy, R.K.; Stahlecker, D.W.; Millsap, B.A.; Jacobson, K.V.; Johnson, A.; Smith, C.S.; Tator, K.J.; Kruse, K.L. Natal dispersal distance of Golden Eagles in the southwestern United States. *J. Fish Wildl. Manag.* **2019**, *10*, 213–218. [CrossRef]
35. Whitfield, D.P.; Fielding, A.H.; Anderson, D.; Benn, S.; Reid, R.; Tingay, R.; Weston, E. Sex difference in natal dispersal distances of Golden Eagles *Aquila chrysaetos* in Scotland. *Ibis* 2023, *in press*.
36. López-López, P. Individual-based tracking systems in ornithology: Welcome to the era of big data. *Ardeola* **2016**, *63*, 103–136. [CrossRef]
37. Hayhow, D.B.; Benn, S.; Stevenson, A.; Stirling-Aird, P.K.; Eaton, M.A. Status of Golden Eagle *Aquila chrysaetos* in Britain. *Bird Study* **2017**, *64*, 281–294. [CrossRef]
38. Whitfield, D.P.; Fielding, A.H.; McLeod, D.R.A.; Haworth, P.F. *A Conservation Framework for Golden Eagles: Implications for Their Conservation and Management in Scotland*; Scottish Natural Heritage Commissioned Report No.193 (ROAME No. F05AC306); SNH: Battleby, UK, 2008.
39. Fielding, A.H.; Haworth, P.; Anderson, D.; Benn, S.; Dennis, R.; Weston, E.; Whitfield, D.P. A simple topographical model to predict Golden Eagle *Aquila chrysaetos* space use during dispersal. *Ibis* **2020**, *162*, 400–415. [CrossRef]
40. Fielding, A.H.; Anderson, D.; Benn, S.; Dennis, R.; Geary, M.; Weston, E.; Whitfield, D.P. Responses of dispersing GPS-tagged Golden Eagles *Aquila chrysaetos* to multiple wind farms across Scotland. *Ibis* **2022**, *164*, 102–117. [CrossRef]
41. Watson, J. *The Golden Eagle*, 2nd ed.; Poyser: London, UK, 2010.
42. Whitfield, D.P.; MacLeod, D.R.A.; Watson, J.; Fielding, A.H.; Haworth, P.F. The association of grouse moor in Scotland with the illegal use of poisons to control predators. *Biol. Conserv.* **2003**, *114*, 157–163. [CrossRef]
43. Whitfield, D.P.; Fielding, A.H.; McLeod, D.R.A.; Haworth, P.F. The effects of persecution on age of breeding and territory occupation in golden eagles in Scotland. *Biol. Conserv.* **2004**, *118*, 249–259. [CrossRef]
44. Whitfield, D.P.; Fielding, A.H.; McLeod, D.R.A.; Haworth, P.F. Modelling the effects of persecution on the population dynamics of golden eagles in Scotland. *Biol. Conserv.* **2004**, *119*, 319–333. [CrossRef]
45. Whitfield, D.P.; Fielding, A.H.; McLeod, D.R.A.; Haworth, P.F.; Watson, J. A conservation framework for the golden eagle in Scotland: Refining condition targets and assessment of constraint influence. *Biol. Conserv.* **2006**, *130*, 465–480. [CrossRef]
46. Whitfield, D.P.; Fielding, A.H.; McLeod, D.R.A.; Morton, K.; Stirling-Aird, P.; Eaton, M.A. Factors constraining the distribution of Golden Eagles *Aquila chrysaetos* in Scotland. *Bird Study* **2007**, *54*, 199–211. [CrossRef]
47. Whitfield, D.P.; Benn, S.; Grant, J.R.; Weston, E. The diet of breeding Golden Eagles in the Central Highlands of Scotland. *Scot. Birds* **2022**, *42*, 202–209.
48. Watson, A.; Wilson, J.D. Seven decades of mountain hare counts show severe declines where high-yield recreational game bird hunting is practised. *J. Appl. Ecol.* **2018**, *55*, 2263–2272. [CrossRef]
49. Watson, J.; Fielding, A.H.; Whitfield, D.P.; Broad, R.A.; Haworth, P.; Nellist, K.; Crane, K.; Macdonald, E.J. Golden eagle breeding performance in relation to climate in western Scotland during the period 1981. In *Birds of Prey in a Changing Environment*; Thompson, D.B.A., Redpath, S., Fielding, A.H., Marquiss, M., Galbraith, C.A., Eds.; The Stationery Office: Edinburgh, UK, 2003; pp. 149–161.
50. Whitfield, D.P.; Reid, R.; Haworth, P.F.; Madders, M.; Marquiss, M.; Tingay, R.; Fielding, A.H. Diet specificity is not associated with increased reproductive performance of Golden Eagles *Aquila chrysaetos* in western Scotland. *Ibis* **2009**, *151*, 255–264. [CrossRef]
51. Fielding, A.H.; Anderson, D.; Benn, S.; Dennis, R.; Geary, M.; Weston, E.; Whitfield, D.P. Non-territorial GPS-tagged golden eagles *Aquila chrysaetos* at two Scottish wind farms: Avoidance influenced by preferred habitat distribution, wind speed and blade motion status. *PLoS ONE* **2021**, *16*, e0254159. [CrossRef] [PubMed]
52. Hoechlin, D.R. Development of golden eaglets in southern California. *West. Birds* **1976**, *7*, 137–152.
53. Peterson, D. *Möt Kungsörnen*; ICA Bokförlag: Västerås, Sweeden, 1997.
54. Phillips, R.A.; Xavier, J.C.; Croxall, J.P. Effects of satellite transmitters on albatrosses and petrels. *Auk* **2003**, *120*, 1082–1090. [CrossRef]
55. Kenward, R.E. *A Manual for Wildlife Radio Tagging*; Academic Press: London, UK, 2001.

56. Sergio, F.; Tavecchia, G.; Tanferna, A.; López Jiménez, L.; Blas, J.; De Stephanis, R.; Marchant, T.A.; Kumar, N.; Hiraldo, F. No effect of satellite tagging on survival, recruitment, longevity, productivity and social dominance of a raptor, and the provisioning and condition of its offspring. *J. Appl. Ecol.* **2015**, *52*, 1665–1675. [CrossRef]
57. Davies, M.; Green, R. A classification of methods used to attach devices to vultures and condors. *Vulture News* **2020**, *78*, 12–17. [CrossRef]
58. Orr-Ewing, D.; Anderson, D.; Weston, E. Thoracic X-Strap Harness. Design and Method. *Vulture News* **2020**, *78a*, 27–30. [CrossRef]
59. Thaxter, C.B.; Ross-Smith, V.H.; Clark, J.A.; Clark, N.A.; Conway, G.J.; Masden, E.; Wade, H.M.; Leat, E.H.K.; Gear, S.C.; Marsh, M.; et al. Contrasting effects of GPS device and harness attachment on adult survival of Lesser Black-backed Gulls *Larus fuscus* and Great Skuas *Stercorarius skua*. *Ibis* **2016**, *158*, 279–290. [CrossRef]
60. García, V.; Iglesias-Lebrija, J.J.; Moreno-Opo, R. Null effects of Garcelon harnessing method and transmitter type on soaring raptors. *Ibis* **2021**, *163*, 899–912. [CrossRef]
61. Anderson, D.; Arkumarev, V.; Bildstein, K.; Botha, A.; Bowden, C.G.R.; Davies, M.; Duriez, O.; Forbes, N.A.; Godino, A.; Green, R.E.; et al. Practical guide to methods for attaching research devices to vultures and condors. *Vulture News* **2020**, *78a*, 1–72, IUCN Species Survival Commission Vulture Specialist Group Technical Publication No.1.
62. Calenge, C. The package adehabitat for the R software package: A tool for the analysis of space and habitat use by animals. *Ecol. Model.* **2006**, *197*, 516–519. [CrossRef]
63. Silva, R.; Afán, I.; Gil, J.A.; Bustamante, J. Seasonal and circadian biases in bird tracking with solar GPS-tags. *PLoS ONE* **2017**, *12*, e0185344. [CrossRef] [PubMed]
64. Thieurmel, B.; Elmarhraoui, A. *Suncalc: Compute Sun Position, Sunlight Phases, Moon Position and Lunar Phase*, R Package Version 0.5.0. 2019.
65. Burnham, K.P.; Anderson, D.R. *Model Selection and Inference: A Practical Information-Theoretic Approach*, 2nd ed.; Springer: New York, NY, USA, 2002.
66. Ens, B.J.; van de Pol, M.; Goss-Custard, J.D. The study of career decisions: Oystercatchers as social prisoners. *Adv. Stud. Behav.* **2014**, *46*, 343–419.
67. Murgatroyd, M.; Roos, S.; Evans, R.; Sansom, A.; Whitfield, D.P.; Sexton, D.; Reid, R.; Grant, J.; Amar, A. Sex-specific patterns of reproductive senescence in a long-lived reintroduced raptor. *J. Appl. Ecol.* **2018**, *87*, 1587–1599. [CrossRef] [PubMed]

diversity

MDPI

Article

Are Roadkills Density-Dependent? Case Study of the Barn Owl (*Tyto alba*)

Kobi Meyrom [1], Reuven Yosef [2] and Motti Charter [3,4,*]

[1] Israel National Barn Owl Project, Society for the Protection of Nature in Israel, HaNegev Street, Tel Aviv 66186, Israel
[2] Eilat Campus, Ben Gurion University of the Negev Eilat Campus, Eilat 8810201, Israel
[3] Shamir Research Institute, Katzrin 1290000, Israel
[4] Department of Geography and Environmental Studies, University of Haifa, Mount Carmel, Haifa 3498838, Israel
* Correspondence: mcharter@geo.haifa.ac.il; Tel.: +972-544-901130

Abstract: Even though birds are some of the most common road-killed animals, it remains to be determined whether avian roadkills are related to breeding numbers and breeding success, mainly due to a lack of study areas that monitor breeding populations and roadkills. We studied whether barn owl breeding numbers and breeding success are related to roadkills. We monitored yearly barn owl breeding numbers (2174 breeding attempts and 1682 adults ringed) and breeding success (9380 nestlings ringed) and monitored 95 km of roads weekly for roadkills from 2009 to 2017 in the Beit Shean and Emek Yizreel Valleys, Israel. During the study period, we documented 1073 road-killed barn owls, of which 328 were ring recoveries. The highest mortality occurred between July to September, coinciding with the barn owl post-fledging period. The number of breeding pairs and the number of nestlings ringed were positively related to the total number of barn owl roadkill, the proportion of roadkill ring recoveries, and the proportion of ring recovered roadkills in the first year of their life. First-year owls represent the majority of ringed owls, accounting for 64.6%, while adult owls compose 35.4%. Notably, a substantial fraction of adult ring recoveries, encompassing 67.2%, may pertain to floaters since we did not observe these individuals as breeding adults. Even though more females were found as roadkill ring recoveries, the proportion of male/female ring recoveries from roadkills was similar to that of adults ringed at the nest boxes. This study is the first that shows that barn owl roadkills are density-dependent and demonstrate the importance of monitoring breeding and population numbers in roadkill studies.

Keywords: roadkill; barn owl; breeding; density; animal–vehicle collisions

Citation: Meyrom, K.; Yosef, R.; Charter, M. Are Roadkills Density-Dependent? Case Study of the Barn Owl (*Tyto alba*). *Diversity* **2023**, *15*, 412. https://doi.org/10.3390/d15030412

Academic Editors: Vladimir D. Dobrev, Dobromir D. Dobrev, Dimitar A. Demerdzhiev and Dimitar Dimitrov

Received: 31 January 2023
Revised: 8 March 2023
Accepted: 9 March 2023
Published: 11 March 2023

1. Introduction

The increasing human population and the accompanying infrastructure growth, such as roads, can significantly impact wildlife populations—for example, by habitat destruction and fragmentation [1,2]. In addition, roads can create dangerous barriers for animals trying to cross from one habitat area to another. The result is a significant increase in roadkill incidents (animal–vehicle collisions) as wildlife are struck by vehicles attempting to cross the road [3].

Roads are the largest source of anthropogenic mortality for many vertebrates [4]. As a result, scientific papers on road ecology have increased annually [5] since it was first described in 1998 [6]. Among the various dangers posed to avian species, vehicular collisions present a grave threat [7]. For example, in the USA alone, 9 to 340 million birds are killed yearly by vehicles [8], millions per year in European countries [9], and 13.8 million in Canada [10].

Various efforts have been made to reduce roadkill, including wildlife crossings, speed limits in sensitive areas, and road design modifications. However, the effectiveness of these measures can vary greatly depending on the location and specific species involved. Numerous studies have tried to explain roadkills by traffic factors, road infrastructure, habitat-related factors, and weather [11]. For example, roadkills were positively related to traffic volumes [12] and landscape connectivity [13,14]. In addition to those factors, it is of high importance to monitor wildlife populations while collecting data on wildlife–vehicle collisions [5] to understand and quantify roadkills' effects on wildlife. However, the relationship between roadkills and wildlife populations may be complex and needs to be fully understood. Therefore, roadkill may lead to changes in population structure and behavior, affecting the persistence of wildlife populations.

Most studies on the effect of population size on roadkills have been conducted in high-income countries, primarily focusing on animal species other than birds [15]. For example, red fox (*Vulpes vulpes*) population densities were related to the number of roadkills [16], as were raccoons [17] (*Procyon lotor*) and rabbits (*Oryctolagus cuniculus*) [18]. In addition, the number of fire salamanders (*Salamandra infraimmaculata*) killed in specific areas was related to the number of roads [19]. Despite the growing recognition of the importance of preserving bird populations, the available research suggests that much more attention is needed to understand the impact of roads on these species and to develop effective conservation strategies.

It is crucial to comprehend the relationship between the number of roadkills, population sizes, and yearly breeding numbers of the species affected. This understanding is crucial to gain insights into the mechanism behind roadkills. Although monitoring roadkills is relatively straightforward compared with monitoring the population demographics of the target species, it is still essential to have baseline data on population numbers. These data provide a foundation to understand the impact of roads on wildlife and allow us to evaluate the evolutionary responses of the species [20] in question to this artificial phenomenon. With these data, it can be easier to fully comprehend the complex interactions between wildlife, roads, and human activity.

Due to several factors, the barn owl (*Tyto alba*) is a common species frequently found as roadkill [21–23]. First, their low-flying flight patterns make them vulnerable to being hit by vehicles. Second, they have a slow reaction time to vehicles [24], contributing to their risk of becoming roadkill. Additionally, their large hunting home ranges [25,26] often overlap with roadsides, where they may prefer hunting [27], increasing their vehicle exposure. These characteristics make barn owls an excellent species for studying roadkill. However, while there have been attempts to investigate the relationship between the occurrence of barn owl populations using broadcast surveys and roadkills [28], these studies have been limited as they lacked data on the actual number of barn owls. This information is crucial as it would clarify the relationship between barn owl populations and the likelihood of roadkills.

In Israel, the use of barn owls as biological pest control agents in agriculture fields is widespread [29,30]. Favorable conditions for hunting and nesting in agricultural areas have led to one of the highest bird population densities in the world. Despite this, the number of breeding pairs and nestlings fledged each year can vary [31], with fluctuations primarily attributed to changes in the rodent population [32] and potentially influenced by weather conditions [33]. Unfortunately, barn owls in Israel also face a significant threat from vehicular-related roadkills as they often live in and around agriculture fields intersected by numerous small roads and highways. Since roadkills are often seasonal, increasing during and after breeding seasons [12,34,35], we expect there will be more barn owl roadkills after breeding, when nestlings fledge and the population is highest. These factors highlight the importance of considering both the benefits and challenges of using barn owls for pest control in agricultural areas.

The present study aimed to investigate the potential relationship between barn owl roadkill incidents, the breeding success of barn owl pairs, the number of breeding pairs, and traffic intensity in Israel. This objective was fulfilled by monitoring barn owl roadkills, breeding pairs, and the number of nestlings. We hypothesize that in years when more barn owl pairs fledged more nestlings, we would expect more barn owls as roadkills. We also studied whether the ringed recoveries from roadkills were related to the breeding numbers and the number of recaptures. We also hypothesized that in years when the barn owls were more productive, there would be more first-year ring recoveries from roadkills.

2. Materials and Methods

The study was conducted (300 km^2) in the adjoining Beit Shean [31] and Emek Yizreel Valleys (32°33′ N, 35°23′ E; Figure 1) and was mainly made up of crop fields comprising cattle fodder (wheat, sweet corn, alfalfa, clover, vetch, and oats), grain crops and seeds (wheat and sweet corn), spices and herbs (oregano, hyssop, basil, and dill), and vegetables (cucumber, pea, etc.) and small villages. In addition, farmers added 606 barn owl nest boxes in and around their fields to increase barn owl populations to be used as biological pest control agents to reduce rodent damage and rodenticide use in their fields [29,30].

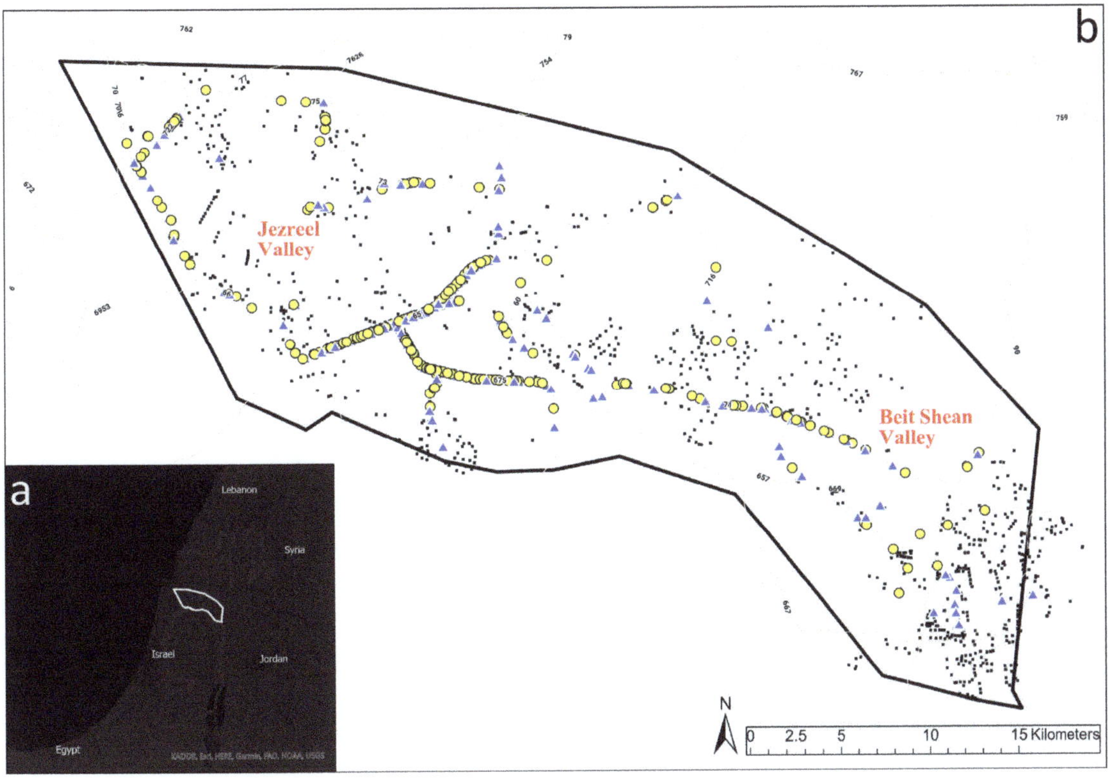

Figure 1. The map on the **bottom left** (**a**) shows the location of the study site in the Beit Shean and Jezreel Valleys in northern Israel and the neighboring countries. The study site is outlined in white on a black background. The larger zoomed-in map (**b**) of the study site with the white background shows the locations of barn owl nest boxes as black dots and the locations of barn owl roadkills collected from 2009 to 2017 as yellow circles (without rings) and blue triangles (with ring recoveries).

2.1. Breeding Data

We monitored, on average, 572 (SE = 19.0) barn owl nest boxes per year. Each nest box was visited two to five times between April 15 and July 30 from 2009 to 2017 to determine the occupied nest boxes. We also calculated the number of fledglings per each laying pair (pairs that laid clutches) by subtracting the number of nestlings when the oldest nestlings were 53 days old, minus any dead nestlings found a week after the fledgling. Based on years of experience researching barn owls in this study system, we assumed a 100% detection probability of breeding pairs, or very close to it, as we have never found any owls breeding outside this period [31,36,37].

All nestlings were ringed yearly [38]. In addition, adult barn owls inside the nest box were also ringed. Males frequently roosted outside the nest box when the nestlings were young, and females sometimes roosted outside the nest boxes when they started to hunt when the oldest nestling was around 25 days [25]. We therefore captured more females than males, and in some cases, neither of the adults were captured. To ensure that we did not harm or disrupt the breeding success of the owls, we took caution when approaching the nest box to avoid flushing them and also when returning the owls after our visit.

2.2. Road Survey

A total of 95 km of roads (road identifying numbers 60, 65, 66, 71, 75, 77, 90, 669, and 6678; Figure 1) were surveyed from two to five times per week by driving slowly to identify roadkills from 2009 to 2017 (n = 9). We only surveyed the roads during the day due to visibility and safety. We used car surveys because compared with foot surveys, car surveys could cover more extensive areas faster and were just as efficient in detecting carcasses on the pavement [39]. During the road survey, we documented each owl's location using the ARC GIS collector mobile data collection app for the ArcGIS platform. The app allowed us to collect, edit, and update geographic data in the field, offline or online, and synchronize changes with a central GIS database. We collected fresh carcasses of owls used in studies on pheomelanin-based coloration and flying strategies [40] and the anatomy of bristles on the nares and rictus of barn owls [41]. In addition to the owls we found during the survey, people would frequently alert us of dead owls in a large WhatsApp group of birders, and we were also alerted by other people, including local rangers. We cleared all other carcasses from the roads to minimize the danger to scavengers and avoid pseudoreplication (counting the same carcass more than once). We defined first-year owls as those in their first calendar year since hatching and adults as those at least one year old. A ring recovery was a ringed dead owl.

2.3. Traffic Intensity

We calculated the traffic intensity from the annual daily traffic collected by the Israel Center Bureau of Statistics (https://teunot.cbs.gov.il/niturtnuaenterprise/ (accessed on 1 January 2023). We presented the traffic data as the mean daily average traffic volume per road (95 km of roads 60, 65, 66, 71, 75, 77, 90, 669, and 6678) during 2009–2017.

2.4. Statistics

The study aimed to investigate the relationships between barn owl roadkills, breeding, ringing data, and traffic intensity over nine years. We used Pearson's correlation to identify correlations between roadkill, breeding, traffic data, and ringing data because the number of observations in our analysis was limited to 9 years, which falls short of the recommended minimum number to carry out a multiple regression analysis. In ecological studies, it is generally advised to have a minimum of 10 to 20 observations per predictor, which can provide a robust estimation of the relationships between variables [42]. In addition, we employed a t-test to determine whether the sex ratio differed between the recovered and recaptured barn owls. For our results, all tests were two-tailed, and p values <0.05 were considered significant. All tests were analyzed using SPSS version 22 software.

3. Results

During the 9-year study, our observations yielded 1073 barn owl roadkills (mean of 119.2 per year, SE = 3.8), among which 328 (30.6%, SE = 7.8) were ring recoveries (Figure 1). The highest mortality occurred during the months of July–September (Figure 2). We monitored 2174 breeding pairs (241.6 adults/year, SE = 23.7) during the study period and ringed 1682 adults (71.6/year, SE = 23.9) and 9380 nestlings (186.9/year, SE = 23.9). The number of breeding pairs and the number of nestlings were positively related (r = 0.93,

$p < 0.001$; Figure 3). The number of breeding pairs ($r = -0.10$, $p = 0.795$; Figure 3) and the number of nestlings ($r = -3.01$, $p = 0.432$; Figure 3) were not related to the year.

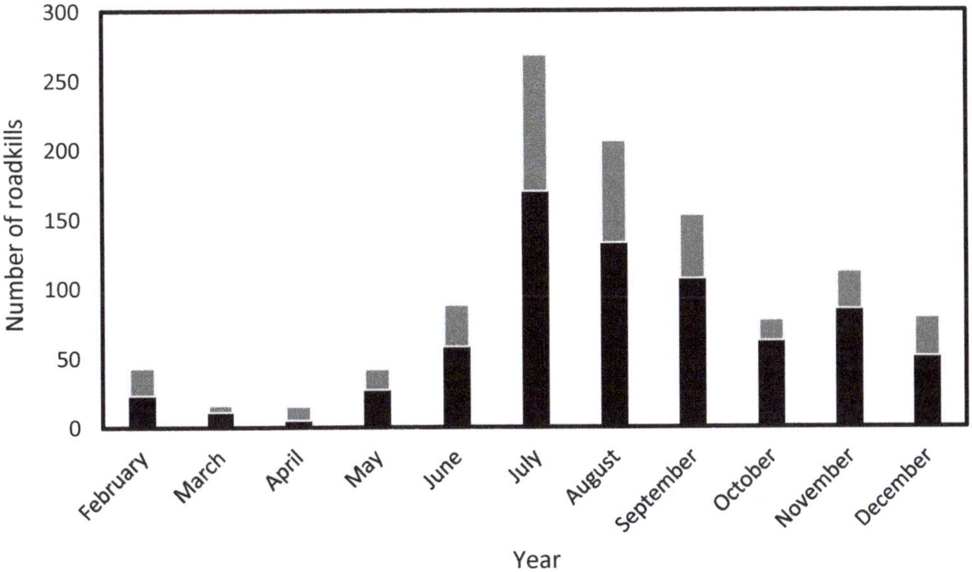

Figure 2. The monthly number of barn owl roadkills that were not ringed (black) and ringed recoveries (gray) from 2009 to 2017 (n = 1073).

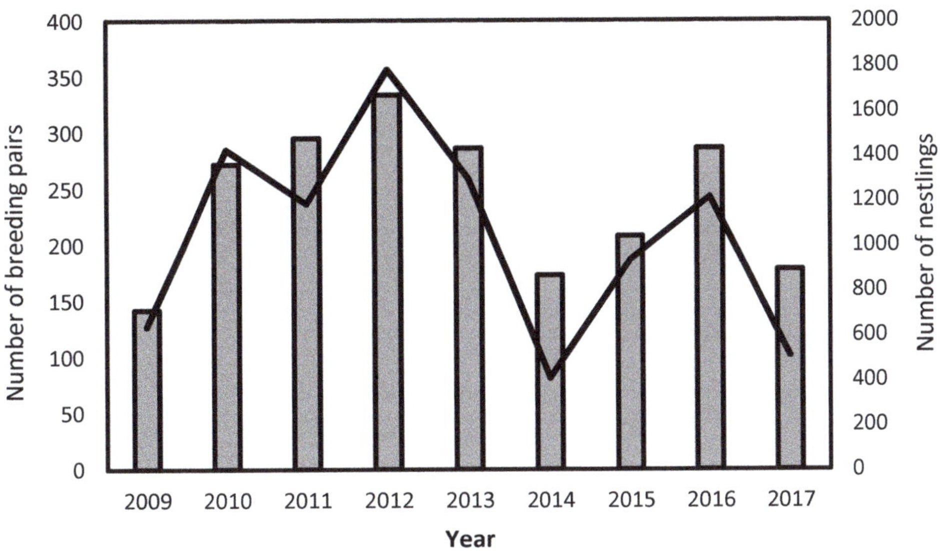

Figure 3. The number of breeding pairs (columns) and the yearly number of nestlings ringed (line) from 2009 to 2017.

The number of barn owl roadkills was positively related to the number of breeding pairs (r = 0.67, p = 0.047; Figure 4a) and the number of nestlings ringed (r = 0.77, p = 0.015; Figure 4b) through the nine years of the study period, whereas it was not related to the traffic intensity (r = −0.32, n = 9, p = 0.398). The proportion of roadkill ring recoveries was positively related to the number of breeding pairs (r = 0.69, n = 9, p = 0.042; Figure 5a) and the number of nestlings ringed (r = 0.69, n = 9, p = 0.038; Figure 5b), whereas it was not related to traffic intensity (r = −0.009, n = 9, p = 0.982). Of the roadkill ring recoveries, adults composed 35.4% (SE = 40.8) and first-year owls 64.6% (SE = 38.5). Of the roadkill adult ring recoveries, 67.2% (SE = 56.6) were owls we ringed as nestlings but never recaptured as breeding adults in active nests. The proportion of roadkill ring recoveries in the first year of their life was related to both the number of breeding pairs (r = 0.80, n = 9, p = 0.010; Figure 6a) and the number of nestlings (r = 0.88, n = 9, p = 0.002; Figure 6b) ringed in the area, whereas it was not related to traffic intensity (r = −0.47, n = 9, p = 0.203). We knew the sex of 89 ring recoveries that we previously captured as breeding adults. The proportion of male/female ring roadkill ring recoveries (29.2% male and 70.8% females) was similar to the proportion of the adults ringed alive at the nest boxes (31.9% male and 68.1% females) (t_{16} = 0.34, p = 0.34).

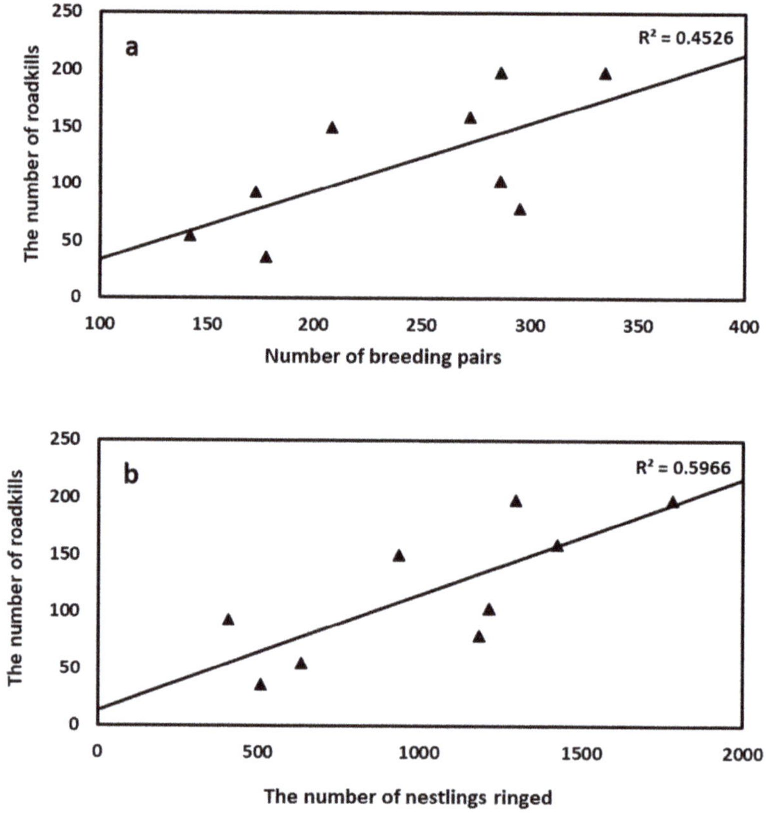

Figure 4. (**a**,**b**) The number of barn owl roadkills from 2009 to 2017 was positively related to the number of barn owl breeding pairs (**a**) and the number of nestlings ringed (**b**).

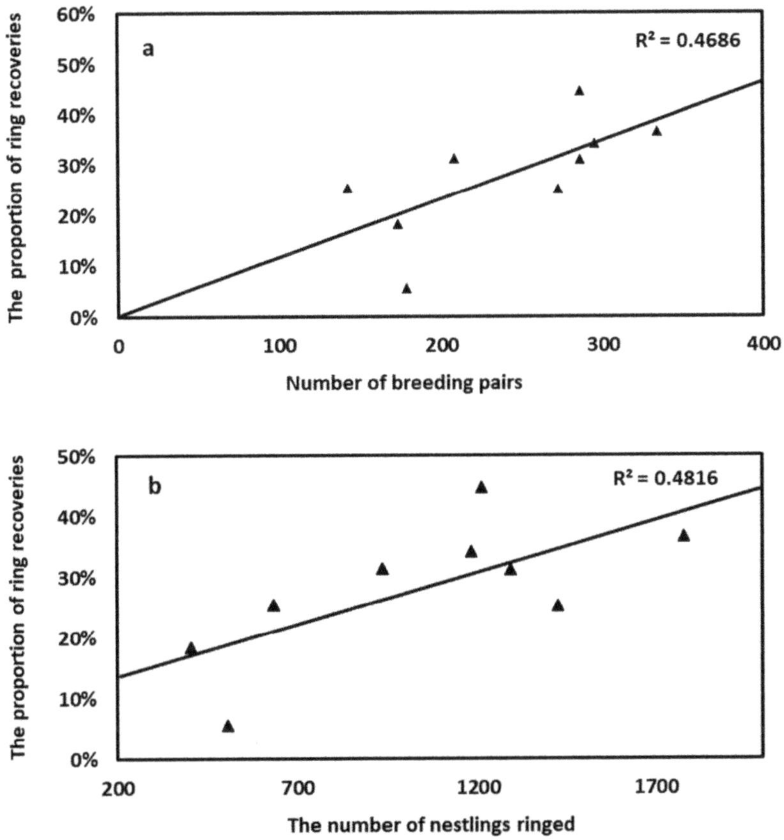

Figure 5. (**a**,**b**) The proportion of barn owl roadkills from 2009 to 2017 that were ring recoveries was positively related to the number of breeding pairs (**a**) and the number of nestlings ringed (**b**).

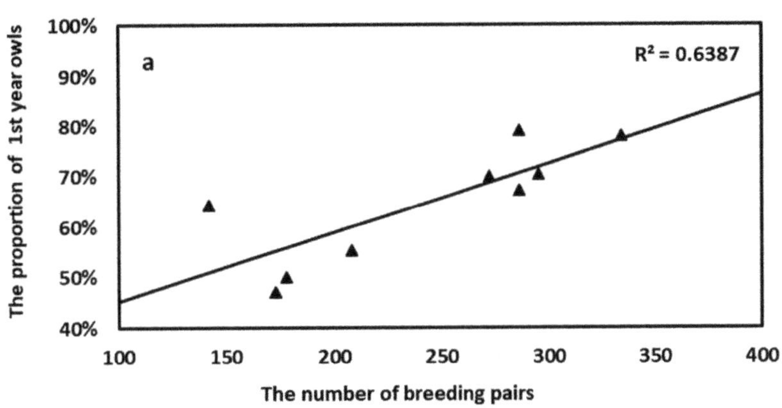

Figure 6. *Cont.*

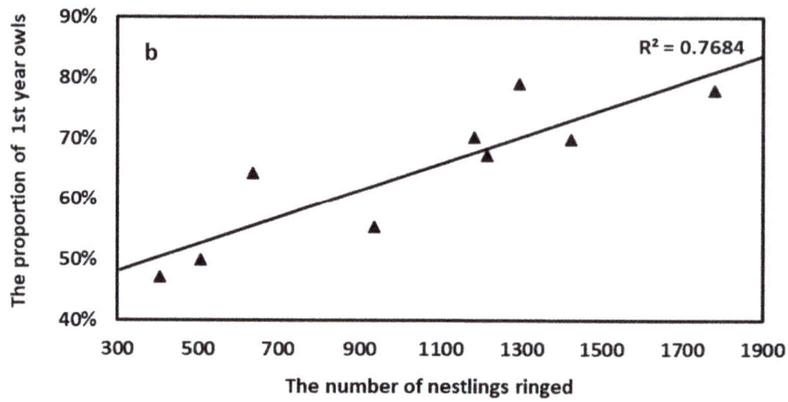

Figure 6. (**a**,**b**) The proportion of barn owl roadkills from 2009 to 2017 that were less than one year old was positively related to the number of breeding pairs (**a**) and the number of nestlings ringed (**b**).

4. Discussion

Understanding population densities and animal behaviors are essential to determine whether mitigation can reduce roadkill [43]. We found that roadkill barn owls were density-dependent. Specifically, the yearly number of barn owl roadkills was directly related to the number of barn owl pairs and the number of nestlings fledged. Furthermore, we found more ring recoveries, and the proportion of first-year owls increased in years with more breeding pairs and more nestling fledge. Therefore, when there are more owls, more are exposed to roads, most likely higher due to intraspecific competition for hunting sites forcing young owls to less attractive habitats [36] and simply because more young owls are dispersing.

Interestingly, compared with the number of owl pairs and nestling, roadkill rates did not fluctuate in response to yearly traffic volume changes, likely because traffic volume remained consistent yearly. Furthermore, in contrast to our investigation, the quantity of roadkill incidents did not correlate with the number of fledged nestlings in 25 km^2 plots in the Netherlands [21]. As barn owl young are capable of fledging far from their nests, a comprehensive evaluation of larger areas, as conducted in our research, may be imperative.

Roadkill monitoring can be used indirectly to monitor wildlife population changes, especially in areas where direct monitoring is impossible. Since roadkills were related to population size, monitoring yearly roadkills can be used to monitor population changes in conservation projects in areas where researchers cannot monitor wildlife demographics. For example, researchers have used roadkill monitoring schemes to determine when species are in decline, as found in the U.K.'s hedgehog (*Erinaceus europaeus*) numbers [44,45]. However, it is essential to validate the relationship between roadkill numbers and population size for each species in each region as different species and regions may have different relationships. In addition, roadkill monitoring should be one of many methods used to monitor wildlife populations because it may vary between populations and is an indirect method that may not always accurately reflect the actual population size. Therefore, it should be used with other methods, such as direct population surveys, to provide a more comprehensive understanding of population changes.

The seasonal peaks in road mortality occurred during the breeding season, especially in the post-fledging period. The temporal mortality of the barn owls, which peaked in July to September, is indicative of both the breeding season and the fledging of the young. When the young barn owls start flying, they will fly far and are at a higher risk of crossing roads. In contrast, adult barn owls are less likely to disperse far from the nest sites after the breeding season, and they also tend to fly less frequently (Charter unpubl. data) after feeding the fledglings. Hence, adult barn owls are primarily at risk while feeding the nestlings.

In comparison with this study, a study in Portugal found the number of barn owl roadkills peaked in November–January after the dispersal period and suggested it was due to the owls' need to fly farther due to a lack of food [46]. Like the barn owls in this study, more Eurasian badgers (*Meles meles*) were road-killed during dispersal [35]. A study in New Zealand found that the three most common mammal and bird species had increased road mortality during dispersal and breeding [12]. Red foxes' (*Vulpes vulpes*) and stone martens' (*Martes foina*) road mortality increased during the breeding season while provisioning young. In the U.K., 13 out of 19 of the most common species showed significant and consistent seasonal variations in road mortality, and many species were also more vulnerable to being road-killed around breeding and dispersal [34].

Juvenile-ringed barn owls in this study were overrepresented in roadkill statistics, with higher proportions of first-year owls (65%) making up more roadkill than adults (35%). This finding is consistent with studies on barn owls conducted in different regions, including the Netherlands (70% juveniles and 30% adults), Florida, USA (61% juveniles and 29% adults) [47], and Idaho, USA (79% juvenile and 21% adults) [22]. The high proportion of juvenile roadkills is likely due to their dispersal behavior as young owls leave their birth sites to establish territories. These movements increase the likelihood of encounters with roads, leading to higher rates of road mortality.

Of the adult recoveries ringed as nestlings, 67.2% we never captured as breeding adults. These adults could be floaters—mature owls that are not breeding. Floater owls are crucial in maintaining the population structure because floaters may also act as potential breeders in case of breeding failures in the population. The elevated count of floaters observed could be attributed to the fact that unlike breeding owls, these birds tend to cover greater distances while flying [48–50], making them more susceptible to road hazards. In addition, floaters may search for prey in less desirable regions farther from occupied nests, such as roads [36].

In this study, the proportion of female/male adults ringed was similar to the sex ratio of roadkills (29% male and 71% female). More females barn owls were also found in roadkills in Florida, USA (26% male and 74% females) [47], but differed from studies in Idaho, USA (42% males and 58 % female) [22], and the Netherlands (50% male and female) [21]. In Israel, females flew farther than males during the breeding season, most likely because males occupied territories first while females were incubating and brooding (Charter unpubl. data), which could increase their exposure to roads and increase the risk of roadkills. More studies using tracking devices are needed to determine whether the movement of males and females may explain exposure to roads and roadkills.

Despite the high frequency of roadkill incidents reported in this study, there were no annual trends or decreases in the size of the barn owl breeding population. It should be noted that roads are not a new phenomenon in the study region, as roadkill incidents were also observed both before and following the duration of this study (Charter pers. observ). It is still unclear whether roadkills limited the population numbers of owls in the study. Fluctuation in the number of breeding pairs and nestlings suggests that barn owl yearly population fluctuations are more likely to be linked to something other than roadkills, such as changes in prey abundance [32,51]. Barn owls in Israel can fledge up to 11 nestlings [31], and the r-selected reproductive strategies and many floater adults seem to offset mortality by roads. Studies on roadkills and breeding in other regions where barn owls raise less young and population density is lower [52], such as in natural habitats, are needed to determine whether less dense and less productive populations may be affected differently. For example, in small populations, the loss of even a few individuals can significantly affect the conservation status of a species. Even though roadkills frequently appear in clusters [53], this may only be the case in some areas, but further studies are needed.

Even though barn owls seem to cross roads/highways regardless of traffic volume [24], there are signs that barn owls prefer nest sites farther away from roads and highways.

For example, in Israel, barn owls preferred to occupy nest boxes farther away from the surrounding [36] roads, most likely due to some disturbance. Similarly, barn owls in Canada were most likely to occupy nest sites in areas with less traffic exposure [54]. Finally, it appears that barn owls are incapable of assessing the speed of vehicles from afar; however, they may try to evade them at the last second by abruptly veering away (Charter unpubl. data). This indicates that barn owls have some level of aversion toward roads and traffic and prefer to breed in quieter, less disturbed areas.

5. Conclusions

This study highlights the importance of monitoring wildlife and roadkills together. In this study, yearly breeding output but not traffic intensity explained the yearly fluctuation in roadkills. Even though many owls were roadkills, the number of pairs did not show a negative trend during the study, and fluctuating breeding numbers occurred before and after this study. The high prey abundance, abundance of nest sites, high breeding success, and a large floater population most likely offset any adverse effects of roadkills in Israel's large barn owl breeding population. There is a need to monitor roadkills and breeding numbers in other smaller populations to determine whether the populations are affected differently. Last, there is a need to ring barn owls to tag them using tracking devices to determine whether there is a difference in movement between breeding and nonbreeding (floaters) adults.

Author Contributions: Conceptualization, M.C. and K.M.; methodology, M.C.; formal analysis, M.C.; data curation, M.C. and K.M.; writing—review and editing, M.C. and R.Y. All authors have read and agreed to the published version of the manuscript.

Funding: This research was funded by the Israel Ministry of Innovation, Science and Technology, grant number 97153.

Institutional Review Board Statement: Bird ringing was conducted using permits from the Israel Nature and Parks Authority.

Informed Consent Statement: Not applicable.

Data Availability Statement: Not applicable.

Acknowledgments: We thank Amir Ezer, Giora Leitner, and Shai Halevi for help in searching for carcasses and Shaul Aviel for ringing. We also thank the Israel National Barn Owl Project of the Society for the Protection of Nature in Israel for logistical assistance. We are grateful to Asaf Aizik (Golan Agri. Innovation Unit) and Society for the Protection of Nature for assistance in GIS. The Israel Ministry of Innovation, Science and Technology funded the study.

Conflicts of Interest: The authors declare no conflict of interest.

References

1. Spellerberg, I.F. Ecological Effects of Roads and Traffic: A Literature Review. *Glob. Ecol. Biogeogr. Lett.* **1998**, *7*, 317–333. [CrossRef]
2. Sánchez-Fernández, M.; Barrigón Morillas, J.M.; Montes González, D.; de Sanjosé Blasco, J.J. Impact of Roads on Environmental Protected Areas: Analysis and Comparison of Metrics for Assessing Habitat Fragmentation. *Land* **2022**, *11*, 1843. [CrossRef]
3. Coffin, A.W. From Roadkill to Road Ecology: A Review of the Ecological Effects of Roads. *J. Transp. Geogr.* **2007**, *15*, 396–406. [CrossRef]
4. Hill, J.E.; DeVault, T.L.; Belant, J.L. Cause-Specific Mortality of the World's Terrestrial Vertebrates. *Glob. Ecol. Biogeogr.* **2019**, *28*, 680–689. [CrossRef]
5. Schwartz, A.L.W.; Shilling, F.M.; Perkins, S.E. The Value of Monitoring Wildlife Roadkill. *Eur. J. Wildl. Res.* **2020**, *66*, 1–12. [CrossRef]
6. Forman, R.T.T.; Alexander, L.E. Roads and Their Major Ecological Effects. *Annu. Rev. Ecol. Syst.* **1998**, *29*, 207–209. [CrossRef]
7. Grilo, C.; Koroleva, E.; Andrášik, R.; Bíl, M.; González-Suárez, M. Roadkill Risk and Population Vulnerability in European Birds and Mammals. *Front. Ecol. Environ.* **2020**, *18*, 323–328. [CrossRef]
8. Loss, S.R.; Will, T.; Marra, P.P. Estimation of Bird-Vehicle Collision Mortality on U.S. Roads. *J. Wildl. Manage.* **2014**, *78*, 763–771. [CrossRef]
9. Erritzoe, J.; Mazgajski, T.D.; Rejt, Ł. Bird Casualties on European Roads — a Review. *Acta Ornithol.* **2003**, *38*, 77–93. [CrossRef]

10. Bishop, C.A.; Brogan, J.M.; Bishop, C.A.; Brogan, J.M. Estimates of Avian Mortality Attributed to Vehicle Collisions in Canada. *Avian Conserv. Ecol.* **2013**, *8*. [CrossRef]
11. Aquino Oddone, A.G.H.E.; Nkomo, S. 'Phumelele L. Spatio-Temporal Patterns and Consequences of Road Kills: A Review. *Animals* **2021**, *11*, 799. [CrossRef]
12. Sadleir, R.M.F.S.; Linklater, W.L. Annual and Seasonal Patterns in Wildlife Road-Kill and Their Relationship with Traffic Density. *New Zeal. J. Zool.* **2016**, *43*, 275–291. [CrossRef]
13. Fabrizio, M.; Di Febbraro, M.; D'Amico, M.; Frate, L.; Roscioni, F.; Loy, A. Habitat Suitability vs Landscape Connectivity Determining Roadkill Risk at a Regional Scale: A Case Study on European Badger (Meles Meles). *Eur. J. Wildl. Res.* **2019**, *65*, 1–10. [CrossRef]
14. Kang, W.; Minor, E.S.; Woo, D.; Lee, D.; Park, C.R. Forest Mammal Roadkills as Related to Habitat Connectivity in Protected Areas. *Biodivers. Conserv.* **2016**, *25*, 2673–2686. [CrossRef]
15. Barrientos, R.; Ascensão, F.; D'Amico, M.; Grilo, C.; Pereira, H.M. The Lost Road: Do Transportation Networks Imperil Wildlife Population Persistence? *Perspect. Ecol. Conserv.* **2021**, *19*, 411–416. [CrossRef]
16. Baker, P.J.; Harris, S.; Robertson, C.P.J.; Saunders, G.; White, P.C.L. Is It Possible to Monitor Mammal Population Changes from Counts of Road Traffic Casualties? An Analysis Using Bristol's Red Foxes Vulpes Vulpes as an Example. *Mamm. Rev.* **2004**, *34*, 115–130. [CrossRef]
17. Gehrt, S.D. Evaluation of Spotlight and Road-Kill Surveys as Indicators of Local Raccoon Abundance. *Wildl. Soc. Bull.* **2002**, *30*, 449–456. [CrossRef]
18. George, L.; Macpherson, J.L.; Balmforth, Z.; Bright, P.W. Using the Dead to Monitor the Living: Can Road Kill Counts Detect Trends in Mammal Abundance? *Appl. Ecol. Environ. Res.* **2011**, *9*, 27–41. [CrossRef]
19. Sinai, I.; Oron, T.; Weil, G.; Sachal, R.; Koplovich, A.; Blaustein, L.; Templeton, A.R.; Blank, L. Estimating the Effects of Road-Kills on the Fire Salamander Population along a River. *J. Nat. Conserv.* **2020**, *58*, 125917. [CrossRef]
20. Brady, S.P.; Richardson, J.L. Road Ecology: Shifting Gears toward Evolutionary Perspectives. *Front. Ecol. Environ.* **2017**, *15*, 91–98. [CrossRef]
21. de Jong, J.; Van Den Burg, A.; Liosi, A. Determinants of Traffic Mortality of Barn Owls (*Tyto alba*) in Friesland, the Netherlands. *Avian Conserv. Ecol.* **2018**, *13*. [CrossRef]
22. Boves, T.J.; Belthoff, J.R. Roadway Mortality of Barn Owls in Idaho, USA. *J. Wildl. Manage.* **2012**, *76*, 1381–1392. [CrossRef]
23. Gomes, L.; Grilo, C.; Silva, C.; Mira, A. Identification Methods and Deterministic Factors of Owl Roadkill Hotspot Locations in Mediterranean Landscapes. *Ecol. Res.* **2009**, *24*, 355–370. [CrossRef]
24. Jacobson, S.L.; Bliss-Ketchum, L.L.; De Rivera, C.E.; Smith, W.P. A Behavior-Based Framework for Assessing Barrier Effects to Wildlife from Vehicle Traffic Volume. *Ecosphere* **2016**, *7*, e01345. [CrossRef]
25. Rozman, G.; Izhaki, I.; Roulin, A.; Charter, M. Movement Ecology, Breeding, Diet, and Roosting Behavior of Barn Owls (*Tyto alba*) in a Transboundary Conflict Region. *Reg. Environ. Chang.* **2021**, *21*, 1–13. [CrossRef]
26. Castañeda, X.A.; Huysman, A.E.; Johnson, M.D. Barn Owls Select Uncultivated Habitats for Hunting in a Winegrape Growing Region of California. *Ornithol. Appl.* **2021**, *123*, 1–15. [CrossRef]
27. Hindmarch, S.; Elliott, J.E.; Mccann, S.; Levesque, P. Habitat Use by Barn Owls across a Rural to Urban Gradient and an Assessment of Stressors Including, Habitat Loss, Rodenticide Exposure and Road Mortality. *Landsc. Urban Plan.* **2017**, *164*, 132–143. [CrossRef]
28. Grilo, C.; Reto, D.; Filipe, J.; Ascensão, F.; Revilla, E. Understanding the Mechanisms behind Road Effects: Linking Occurrence with Road Mortality in Owls. *Anim. Conserv.* **2014**, *17*, 555–564. [CrossRef]
29. Peleg, O.; Nir, S.; Meyrom, K.; Aviel, S.; Roulin, A.; Izhaki, I.; Leshem, Y.; Charter, M. Three Decades of Satisfied Israeli Farmers: Barn Owls (*Tyto alba*) as Biological Pest Control of Rodents. In Proceedings of the 28th Vertebrate Pest Conference, Rohnert Park, CA, USA, 26 February–1 March 2018; Woods, D.M., Ed.; University of California Davis: Davis, CA, USA, 2018; pp. 194–203.
30. Meyrom, K.; Motro, Y.; Leshem, Y.; Aviel, S.; Izhaki, I.; Argyle, F.; Charter, M. Nest-Box Use by the Barn Owl Tyto alba in a Biological Pest Control Program in the Beit She'an Valley, Israel. *Ardea* **2009**, *97*, 463–467. [CrossRef]
31. Charter, M.; Rozman, G. The Importance of Nest Box Placement for Barn Owls (*Tyto alba*). *Animals* **2022**, *12*, 2815. [CrossRef] [PubMed]
32. Charter, M.; Izhaki, I.; Roulin, A. The Relationship between Intra–Guild Diet Overlap and Breeding in Owls in Israel. *Popul. Ecol.* **2018**, *60*, 397–403. [CrossRef]
33. Charter, M.; Izhaki, I.; Meyrom, K.; Aviel, S.; Leshem, Y.; Roulin, A. The Relationship between Weather and Reproduction of the Barn Owl Tyto alba in a Semi-Arid Agricultural Landscape in Israel. *Avian Biol. Res.* **2017**, *10*, 253–258. [CrossRef]
34. Raymond, S.; Schwartz, A.L.W.; Thomas, R.J.; Chadwick, E.; Perkins, S.E. Temporal Patterns of Wildlife Roadkill in the UK. *PLoS One* **2021**, *16*. [CrossRef] [PubMed]
35. Grilo, C.; Bissonette, J.A.; Santos-Reis, M. Spatial-Temporal Patterns in Mediterranean Carnivore Road Casualties: Consequences for Mitigation. *Biol. Conserv.* **2009**, *142*, 301–313. [CrossRef]
36. Charter, M.; Leshem, Y.; Meyrom, K.; Peleg, O.; Roulin, A. The Importance of Micro-Habitat in the Breeding of Barn Owls *Tyto alba*. *Bird Study* **2012**, *59*, 368–371. [CrossRef]
37. Charter, M.; Izhaki, I.; Leshem, Y. Effects of the Risk of Competition and Predation on Large Secondary Cavity Breeders. *J. Ornithol.* **2010**, *151*, 791–795. [CrossRef]

38. Charter, M.; Meyrom, K.; Leshem, Y.; Aviel, S.; Izhaki, I.; Motro, Y. Does Nest Box Location and Orientation Affect Occupation Rate and Breeding Success of Barn Owls *Tyto alba* in a Semi-Arid Environment? *Acta Ornithol.* **2010**, *45*, 115–119. [CrossRef]
39. Guinard, É.; Julliard, R.; Barbraud, C. Motorways and Bird Traffic Casualties: Carcasses Surveys and Scavenging Bias. *Biol. Conserv.* **2012**, *147*, 40–51. [CrossRef]
40. Charter, M.; Leshem, Y.; Izhaki, I.; Roulin, A. Pheomelanin-Based Colouration Is Correlated with Indices of Flying Strategies in the Barn Owl. *J. Ornithol.* **2015**, *156*, 309–312. [CrossRef]
41. Delaunay, M.G.; Charter, M.; Grant, R.A. Anatomy of Bristles on the Nares and Rictus of Western Barn Owls (*Tyto alba*). *J. Anat.* **2022**, *241*, 527–534. [CrossRef] [PubMed]
42. Gotelli, N.J.; Ellison, A.M. *A Primer of Ecological Statistics*; Sinauer Associates: Sunderland, MA, USA, 2004; Volume 1.
43. Rytwinski, T.; Soanes, K.; Jaeger, J.A.G.; Fahrig, L.; Findlay, C.S.; Houlahan, J.; Van Ree, R.D.; Van Der Grift, E.A. How Effective Is Road Mitigation at Reducing Road-Kill? A Meta-Analysis. *PLoS One* **2016**, *11*, e0166941. [CrossRef] [PubMed]
44. Wembridge, D.E.; Newman, M.R.; Bright, P.W.; Morris, P.A. An Estimate of the Annual Number of Hedgehog (Erinaceus Europaeus) Road Casualties in Great Britain. *Mammal Commun.* **2016**, *2*, 8–14.
45. Pettett, C.E.; Johnson, P.J.; Moorhouse, T.P.; Macdonald, D.W. National Predictors of Hedgehog Erinaceus Europaeus Distribution and Decline in Britain. *Mamm. Rev.* **2018**, *48*, 1–6. [CrossRef]
46. Borda-de-Água, L.; Grilo, C.; Pereira, H.M. Modeling the Impact of Road Mortality on Barn Owl (*Tyto alba*) Populations Using Age-Structured Models. *Ecol. Modell.* **2014**, *276*, 29–37. [CrossRef]
47. Moore, T.G.; Mangel, M. Traffic Related Mortality and the Effects on Local Populations of Barn Owls (*Tyto alba*). In *Trends in Addressing Transportation Related Wildlife Mortality, Proceedings of the Transportation Related Wildlife Mortality Seminar, Orlando, FL, USA, 30 April–2 May 1996*; Evink, G.L., Garrett, P., Berry, J., Eds.; Florida Department of Transportation: Tallahassee, FL, USA, 1996; pp. 58–96.
48. Rohner, C. Great Horned Owls and Snowshoe Hares: What Causes the Time Lag in the Numerical Response of Predators to Cyclic Prey? *Oikos* **1995**, *74*, 61–68. [CrossRef]
49. Penteriani, V.; Ferrer, M.; Delgado, M.M. Floater Strategies and Dynamics in Birds, and Their Importance in Conservation Biology: Towards an Understanding of Nonbreeders in Avian Populations. *Anim. Conserv.* **2011**, *14*, 233–241. [CrossRef]
50. Van Eeden, R.; Whitfield, D.P.; Botha, A.; Amar, A. Ranging Behaviour and Habitat Preferences of the Martial Eagle: Implications for the Conservation of a Declining Apex Predator. *PLoS One* **2017**, *12*, e0173956. [CrossRef] [PubMed]
51. Klok, C.; De Roos, A.M. Effects of Vole Fluctuations on the Population Dynamics of the Barn Owl *Tyto alba*. *Acta Biotheor.* **2007**, *55*, 227–241. [CrossRef]
52. Roulin, A. Barn Owls: Evolution and Ecology. *Barn Owls* **2020**. [CrossRef]
53. Bartonička, T.; Andrášik, R.; Duľa, M.; Sedoník, J.; Bíl, M. Identification of Local Factors Causing Clustering of Animal-Vehicle Collisions. *J. Wildl. Manage.* **2018**, *82*, 940–947. [CrossRef]
54. Hindmarch, S.; Krebs, E.A.; Elliott, J.E.; Green, D.J. Do Landscape Features Predict the Presence of Barn Owls in a Changing Agricultural Landscape? *Landsc. Urban Plan.* **2012**, *107*, 255–262. [CrossRef]

diversity

MDPI

Article

Selection of Nesting Habitat and Insular Niche Separation of Two Sympatric Aquila Species

Maria Solanou [1,2], Eirini Trypidaki [1], Elisavet Georgopoulou [1], Konstantinos Damianakis [1], Afroditi Kardamaki [1] and Stavros M. Xirouchakis [1,*]

1 Natural History Museum of Crete, School of Sciences & Engineering, University of Crete, University campus (Knossos), 71409 Heraklion, Greece

2 Institute of Marine Biological Resources and Inland Waters (IMBRIW), Hellenic Centre for Marine Research (HCMR), Former US Base at Gournes, P.O. Box 2214, 71003 Heraklion, Greece

* Correspondence: sxirouch@nhmc.uoc.gr

Abstract: *Aquila chrysaetos* and *Aquila fasciata* are two congeneric eagle species distributed in the Mediterranean region which are supposed to compete for similar breeding and foraging resources. In the present study, bioclimatic, topographic, and human-related habitat parameters were investigated for 64 and 75 nest sites of Golden and Bonelli's eagles, respectively. The nests were found during fieldwork undertaken from 1995–2020. Overall, the habitat parameters that best discriminated nest site selection were associated with elevation, temperature, and land use with topographic variables being most powerful for niche separation. Univariate analysis, regression, and species distribution modeling identified a strong association of the species with altitude pinpointing the Golden eagle's mountainous and continental character and the Bonelli's eagle being a lowland and coastal species. Golden eagle nests were situated away from human settlements on steep cliffs in higher altitude areas with transitional woodland-shrub vegetation. In contrast Bonelli's eagle nests were located on low-altitude warmer zones, closer to the coast and human settlements with more natural grasslands in their vicinity. The ecological niche separation of the two species was best described by altitude and temperature, though no clear-cut evidence was detected for their competitive exclusion. Inter-specific nearest neighboring distance was found statistically significant only for the Golden eagle which seems to be less tolerant in its co-existence with the Bonelli's eagle. Conservation measures for both species should target territories under human pressure, though more research should focus on the species range use and habitat heterogeneity within overlapping territories.

Keywords: *Aquila chrysaetos*; *Aquila fasciata*; maximum entropy; logistic regression; Crete; insular environment; niche similarity; nest habitat

Citation: Solanou, M.; Trypidaki, E.; Georgopoulou, E.; Damianakis, K.; Kardamaki, A.; Xirouchakis, S.M. Selection of Nesting Habitat and Insular Niche Separation of Two Sympatric Aquila Species. *Diversity* **2023**, *14*, 1136. https://doi.org/10.3390/d14121136

Academic Editors: Vladimir D. Dobrev, Dobromir D. Dobrev and Dimitar A. Demerdzhiev

Received: 13 November 2022
Accepted: 15 December 2022
Published: 18 December 2022

1. Introduction

Eagles are long-lived raptors at the top of the food chain occupying sizeable home ranges in order to meet their ecological requirements; a fact the makes their conservation and management quite difficult particularly within human-modified landscapes [1,2]. Aquila species are quite vulnerable to land use changes, rural and urban development, and infrastructure facilities, all of which result in the loss of suitable habitat affecting their population dynamics [3–5]. However, apart from human induced factors a major source of habitat exclusion might be the competitive interactions among species with similar ecological niches and overlapping distributions [6–9]. In the Mediterranean region this is the case of the Golden (*Aquila chrysaetos*) and the Bonelli's eagle (*Aquila fasciata*) who share similar nesting and foraging habitat [10,11] with potential constraints on the breeding density of the latter [12–15]. Interspecific competition has been taken into account in many conservation efforts for the recovery of the Bonelli's eagle in Europe as many of the species deserted territories have not been recolonized due to their usurpation by Golden eagles [14–17].

In Greece, the Golden eagle is mainly distributed on the mainland and the island of Crete with an estimated population of 100–150 breeding pairs [18–20]. In contrast the Bonelli's eagle is restricted to southern continental Greece (e.g., Peloponnese) and the Aegean islands including Crete [18] with a population numbering no more than 140 breeding pairs. Both species are enlisted in the Greek Red Data Book as "Endangered" and "Vulnerable", respectively, due to negative population trends that have been mainly caused by human persecution, secondary poisoning, collision with energy infrastructure and land use changes [20].The island of Crete harbors viable and relatively stable populations of the species [21,22] and along with Sicily constitutes the only Mediterranean island where they co-exist [23]. Both species are cliff nesting and depend on similar prey, i.e., medium to small size mammals and birds [20,21].

According to the theory of competitive exclusion, sympatric species with similar life-history and ecological traits would exhibit some form of distinction in ecological niche and resource partitioning in space or time or dietary segregation [24–28]. In this context and given that island raptors occupy niches of wider breadth compared to their mainland counterparts [29–31], Crete is a good model area to quantify habitat features that facilitate niche separation of the two species. Moreover, understanding the processes undergoing their spatial distribution on the island and their demand on natural resources are pivotal for supporting effective management policies for both species at a national level.

In the aforementioned framework the objectives of the current study were to: (a) census the eagle territories on the island of Crete giving special emphasis on defining the overlapping ones, (b) investigate the environmental and anthropogenic factors that influence nest site selection or limit territory occupancy, (c) asses niche similarity and partitioning within eagle breeding range, and (d) construct habitat suitability maps for depicting potential nesting areas for recolonization. The ultimate aim was to elucidate the ecological mechanism for the coexistence of the Golden and the Bonelli's eagles on Crete and provide baseline guidance for the efficient conservation planning and simultaneous management for their populations.

2. Materials and Methods

2.1. Study Area

Crete is the largest island of Greece, with an area of 8261 km^2 and a length from east to west of 260 km. Its width ranges between 12 and 60 km and its terrain is characterized by mountainous areas (60%) that are crisscrossed by cracks, numerous gorges (>120), rocky outcrops, and steep cliffs [32]. The three major massifs extending from west to east are the Lefka Ori mountain range (2450 m), Mount Psiloritis (2457 m), and Mount Dikti (2100 m). The island's overall terrain consists of some plains and many hills covered with extensive vineyards and olive groves. Depending on the longitude and the altitude, the annual rainfall ranges between 400 and 2000 mm extending over 70–130 rainy days per year. The average temperature in winter ranges between 11–16 °C while in summer between 21–28 °C [33]. Winters are mild and usually wet, while summers are dry and hot. Snow is common in the mountains, and can persist from late November until early May. Despite having been altered by humans for at least 3000 years, the predominant vegetation is dry-leafed maquis and the cushiony shrubs, i.e., phrygana [34].

2.2. Fieldwork

The study area, namely the entire island of Crete, was divided to 330 sample units of 5 × 5 km, namely quadrats of 25 km^2. Nest searches for both eagle species were conducted systematically from 1995 to 1999 in 165 quadrats covering 50% of the island's surface. Fieldwork constituted of midday surveys carried out during mid-November–late March and observations were made from a vehicle moving at a mean speed of 35 km per hour [35]. Fieldwork usually started two hours after sunrise and ended at 16:00 pm, although mean effective time was 6.5 h per day. All eagles observed with unaided eye on both sides of the road were recorded and identified by two observers [36–38]. For each sighting, eagle position was recorded on a 1:50,000 scale map (Hellenic Military Geographical Service) and notes were kept on date, time, locality, altitude, number of individuals, and activity codes [1,35,39]. Additional random surveys were undertaken over the entire island during 2000–2005 and 2015–2020. Territory occupancy was validated by observations made from vantage points in areas with suitable breeding habitat and high eagle use. Cliffs were inspected during the pre-breeding period looking for eagles performing undulating flights, carrying nest material or defending a nesting territory. All eagle nests were found by the use of 10 × 50 binoculars and a 30–60× spotting scope and their geographical position was fixed on a handheld hiking GPS device (Garmin Montana 680 t) on a WGS 84 coordinate reference system. Their position was transferred, digitized, and transformed into UTM coordinates of the Hellenic Geodetic Reference System (Hellenic Mapping & Cadastral Organization, 1995) by the aid of a Geographic Information System (i.e., ArcGIS Pro, ESRI 2021).

2.3. *Data Collection & Processing*

Environmental and human pressure variables were selected according to their relevance with the species nesting ecology (Table 1). A subset of current climatic variables was designated with regard to the species breeding seasons (January–March) and were obtained from the open access WorldClim database at a resolution of 30 s (https://www.worldclim.org/, accessed on 20 July 2021). Human-related variables were calculated using satellite imagery (http://geodata.gov.gr/; https://www.openstreetmap.org, accessed on 20 July 2021), while topographic variables were produced by the aid of a Digital Elevation Model (DEM-Digital Elevation Model) of Crete with a 50 × 50 m^2 resolution (http://srtm.csi.cgiar.org/, accessed on 20 July 2021). The circular variable of "aspect" (0–360°) was transformed into two linear ones, namely "eastness" and "northness", using cosine and sine, respectively, and their measurements ranged between −1 and 1, with positive values indicating an eastward and northward tilt, correspondingly. The normalized enhanced vegetation index (EVI) [39] was used to represent plant biomass and data for two periods, i.e., Winter (December–February) and Summer (May–July), were extracted from MODIS satellite images at 16 day interval and 250 m resolution. Average EVI values for the years 2000 to 2020 were calculated by the R library MODIStsp [40] with the R 4.1.1 software [41]. Concerning land use variables, the Corine Land Cover database was used covering the years 1980–2000 and 2000–2020 [42,43] in accordance to the main fieldwork data collection periods (https://land.copernicus.eu/pan-european/corine-land-cover, accessed on 20 July 2021). Land cover types were collapsed to the most meaningful ones in biological terms for eagle territories (Table 1) and their percentages were calculated in buffers of 6 km and 4 km around Golden and Bonelli's eagle nests, respectively. These figures were regarded as the shortest foraging radius of a breeding eagle pair within a hypothetical circular territory and an average home range size of ca. 100 km^2 for the Golden eagle and 50 km^2 for the Bonelli's eagle [10,21,44–49]. The latter approach was considered to be representative for central place foragers when breeding such as the study species [50] and consistent to the radius of major activity around nest sites reported in the literature [15,51,52]. As a proxy for eagle competition, the nearest neighboring inter-specific Euclidean (horizontal) distance (NNID) was calculated, and was further used in the statistical analysis.

Table 1. Environmental data collected for *Aquila chrysaetos* and *Aquila fasciata* nesting territories on Crete and used as explanatory variables for species distribution modeling and ecological niche separation.

Variables	Description
Climate	Temperature Seasonality (BIO4): Temperature monthly variation (CV) * Temperature Annual Range (BIO7): Temperature annual variation (°C) Winter temperature (BIO11): Mean Temperature of Coldest Quarter (°C) * Precipitation Seasonality (BIO15): Precipitation monthly variation (CV) Winter rainfall (BIO19): Precipitation of Coldest Quarter (mm) *
Topography	Elevation (m) Slope (°) DistSea: Distance to coastline (m) Elevation (m) Northness (degrees) Eastness (degrees)
Vegetation cover	EVI winter: EVI December-February EVI summer: EVI May-July
Land Cover	Bare land Forests Agriculture Natural grassland Transitional woodland-shrubland Sclerophyllous Olive groves Water bodies Pastures Human infrastructures
Competition	NNID (m)
Human pressure	DistTown: Distance to nearest town (≥10,000 inhabitants) (m) DistSettlements: Distance to nearest settlement (≤10,000 inhabitants) (m) DistRoad: Distance to nearest road (m)

* Variables excluded from the GLM analysis due to multicollinearity.

2.4. Statistical Analysis

2.4.1. Nesting Habitat & Territory Overlap

All explanatory variables were extracted for individual eagle nest sites. Univariate comparisons were made by the use of Student's test at a 0.05 level of statistical significance or two-tailed Mann–Whitney–Wilcoxon test in case the assumption of normality of data distribution was violated [53]. Territory overlap was delineated by using the Predicting Aquila Territory (PAT) model which provides an estimate of home range use based on nest site locations, elevation and terrain [51,54–56]. Major assumptions of the model were the excessive use of ridges and rocky outcrops around nest sites and the complete avoidance of human activity areas (i.e., roads and settlements) and marine or freshwater bodies. Relevant spatial data (i.e., polygons) were downloaded from open access databases (http://geodata.gov.gr/, https://www.openstreetmap.org, accessed on 25 July 2021), or were constructed by using a digital elevation model (DEM) of Crete (http://srtm.csi.cgiar.org/, accessed on 25 July 2021). For eagle territories with multiple alternative nests [57], their geometric mean was taken into account. Thiessen polygons were used to establish territorial boundaries between nests, while a maximum radius of 6 km and 4 km were defined around Golden and Bonelli's nests, respectively, in case of absence of neighboring territories [55,58–60]. The grid cells constructed in the ArcGIS Pro environment were given certain values and predictive maps with areas of high eagle use within each territory were produced.

2.4.2. Niche Similarity & Partitioning

An equivalency test was implemented in order to address whether the congeneric eagle species are effectively identical in their predicted realized environmental distributions [61]. At first, two similarity metrics, namely the Schoener's *D* [62] and the Warren's *I*, were calculated by the ENMTools R library [63] both ranging from 0 (no overlap) to 1 (niche models identical). Subsequently, all occurrence data from both species were pooled into one dataset and were tested by generating new overlapping values via a repeatedly randomization process of one hundred (100) permutations [64]. Empirical overlapping values were finally compared to the critical ones in order to assess the species' overlap. Namely, empirical values lower than the corresponding empirical ones are indicative of a relatively low overlap and important niche differentiation.

2.4.3. Nest Site Selection & Habitat Suitability

Nest site locations for each species were pooled for the entire study period and filtered by the aid of the spThin R library [65] so as to retain sites at least 1 km apart and avoid spatial autocorrelation issues. For the explanatory variables, two subsets of data were formulated: one consisting by the landcover types and another one with the rest of the environmental and human pressure variables. All variable layers were processed as regular grids at a spatial resolution of 400 × 400 m^2 in ArcGIS Pro (ESRI 2021). A logistic regression model was built for each set of explanatory variables, assuming a binomial error distribution and a logit link function for binary response variables (i.e., 1 = nest presence, 0 = nest absence). Absence points were randomly generated at a minimum distance of 1 km apart, excluding points falling at a distance of 3 and 2 km from Golden and Bonelli's nests, which correspond to their mean core area radius, respectively [15,51,52]. Multicollinearity between explanatory variables (Table 1) was checked by the Variance Inflation Factor (VIF), where only those with values smaller than 10 were retained in the dataset so to avoid overparameterization [53,66–68]. Missing values were removed, and the top model's performance was assessed using k-fold cross validation with five folds [69].

Furthermore, a Maximum Entropy method was applied using the MaxEnt ver. 3.4.4 software [70] by using presence-only data and default settings in order to achieve repeatability. NNID were excluded from the analysis as it was impossible to incorporate pseudoabsence data for them. Overall, 5000 background points were generated for each species and a 5-fold cross-validation process was applied by randomly selecting 75% of them as training and the other 25% as testing data [69]. The technique's prediction accuracy was evaluated by area under the receiver operating characteristic curve (AUC) metric [71,72]. Values for the parameter indicate high accuracy with a perfect discriminatory ability (i.e., >0.9), moderate accuracy (0.70–0.90), or low accuracy of the model with discrimination no better than random (ca. 0.5) [73]. Finally, the predictors' importance was evaluated with jackknife analysis and a logistic output was selected because of its extensive use in ecological research and ease of interpretation. In addition, the overlap of the of the habitat suitability maps was checked by converting the probability surface to presence–absence by using the Maximum of sensitivity plus specificity optimalization criterion [70]. This threshold minimizes the mean of error rate for positive and negative observations and determines the optimum cut-off point in the ROC curve and produces binary maps.

3. Results

3.1. Nesting Habitat & Territory Overlap

Overall, 64 and 75 nest sites belonging to Golden and Bonelli's eagles, respectively, were found on precipitous cliffs. Statistically significant differences were detected between climatic, topographic, and human-related parameters, which described the species nesting sites (Table 2). More specifically, the nesting cliffs of Golden eagles were located in open areas on higher altitude with intermediate woodland-shrub vegetation in their surroundings and received more rainfalls during the breeding season. In addition, the species was found breeding away from human settlements and warm coastal areas, though being more tolerant to temperature variability. In contrast, the Bonelli's eagle nest sites were located on cliffs of lower altitude in warmer zones, closer to human settlements and the coast, which is covered by various vegetation types and receive irregular rainfalls. However, some of the aforementioned differences became redundant when 13 Golden and 19 Bonelli's eagle nests (alternative nests included) in nine overlapping territories were examined (Table 3).

Table 2. Mean values ± SD and range (min-max) of environmental and habitat parameters of univariate analysis of Golden and the Bonelli's eagle nest sites in Crete. (*Significant p values in bold and italic*).

	Aquila chrysaetos (n = 64)		*Aquila fasciata (n = 75)*		
Variables	*Mean ± SD*	*Range*	*Mean ± SD*	*Range*	*p*
Temperature Seasonality	574.4 ± 15.8	541.5–605.43	566.4 ± 14.23	528.5–597.5	***0.01***
Temperature Annual Range	22.1 ± 0.7	20.2–23.4	22 ± 0.77	19.3–23.1	0.62
Winter temperature	8.783 ± 2	4–11.82	10.3 ± 1.31	6.75–12.25	***0.0001***
Precipitation Seasonality	84.8 ± 2.1	81.1–89.5	85.7 ± 1.49	82.8–90.4	***0.007***
Winter rainfall	386.2 ± 73	278–544	343.2 ± 61.6	251–526	***0.002***
Elevation	699.7 ± 329.3	152.5–1378	396 ± 243	56–983	***3.388×10^{-6}***
Slope	25.4 ± 9.9	8.15–45.2	25.47 ± 10.9	4.7–47.8	0.97
EVI summer	0.21 ± 0.04	0.13–0.31	0.21 ± 0.05	0.13–0.34	0.64
EVI winter	0.19 ± 0.05	0.1–0.27	0.22 ± 0.05	0.11–0.33	***0.0025***
DistTown	26152 ± 9792	5536–45907	24544 ± 11796	4327–51842	0.46
DistSettlements	2180 ± 1633	707–11037	1640 ± 1602	141–10108	***0.0009***
DistRoad	444 ± 431	0–1887	292 ± 254	0–10059	0.14
DistSea	6807 ± 4399	721–16505	4271 ± 4149	300–19731	***0.0009***
Sclerophyllous	0.29 ± 0.22	0–0.88	0.24 ± 0.24	0–0.8	0.1
Olive groves	0.06 ± 0.07	0–0.34	0.09 ± 0.13	0–0.63	0.73
Bare land	0.08 ± 0.11	0–0.49	0.11 ± 0.2	0–0.78	0.18
Forests	0.09 ± 0.16	0–0.72	0.07 ± 0.14	0–0.76	0.19
Agriculture	0.06 ± 0.06	0–0.21	0.08 ± 0.09	0–0.33	0.69
Natural grassland	0.32 ± 0.21	0–0.76	0.36 ± 0.22	0–0.79	0.36
Woodland-shrub	0.08 ± 0.1	0–033	0.05 ± 0.09	0–0.27	***0.007***

Table 3. Mean values ± SD and range (min-max) of significantly different environmental and habitat parameters of Golden and the Bonelli's eagle nest sites on overlapping territories in Crete.

	Aquila chrysaetos		*Aquila fasciata*		
Variables	*Mean ± SD*	*Range*	*Mean ± SD*	*Range*	*p*
Elevation	812 ± 254	435–1264	546 ± 242	200–983	0.005
DistS	2285 ± 1400	707–5855	1031 ± 574	300–2730	0.002
DistR	530 ± 604	0–1887	317 ± 270	0–860	0.04
Olive groves	0.05 ± 0.08	0–0.46	0.15 ± 0.12	0–0.36	0.005
Bare land	0.10 ± 0.14	0–0.31	0.08 ± 0.20	0–0.74	0.02

In fact, three variables were retained, namely the altitude of the nests and their distance from the nearest road and human settlement, showing that the Bonelli's eagle might breed closer to urban areas. In addition, two habitat variables distinguished furthermore the two eagle species with the Golden eagle nesting in more open areas with sparse vegetation and the Bonelli's eagle in areas covered by a higher proportion of cultivations namely olive groves. This pattern was also depicted geographically in the species distribution where they seemed to co-exist in a transitional zone from mountainous semi-natural habitats to human-altered middle lands with extensive farming at their periphery.

Both similarity indices between the nesting habitats of the two species were lower than expected by random as shown by their calculated empirical and permuted critical values. More specifically the D and I metrics were significantly different than their empirical values (*A. chrysaetos*: 0.80 vs. 0.53 and *A. fasciata*: 0.90 vs. 0.53) implying that the nesting habitat of the two eagle species are statistically different, even if both of them select similar breeding habitats or nest close to each other. Likewise, the PAT model (Figure 1) showed that grids of high eagle use in overlapping territories accounted for 25% and 44% of those belonging to Golden and Bonelli's eagles, respectively.

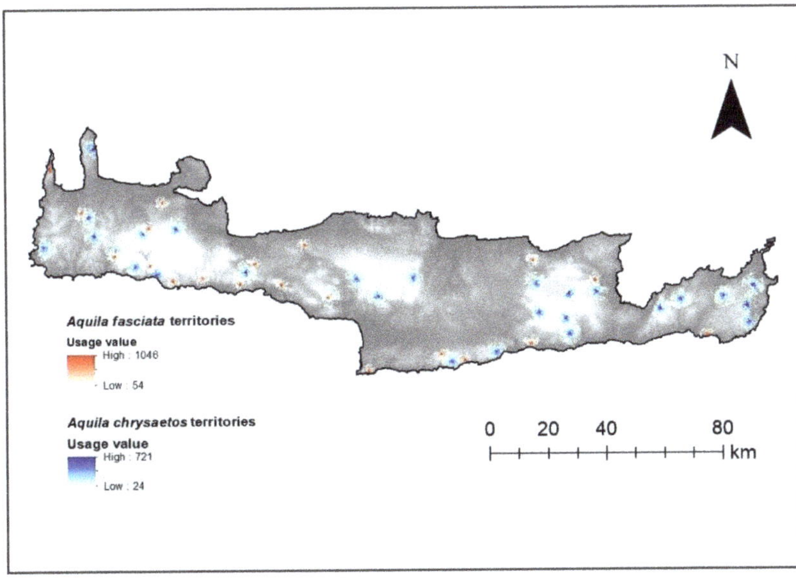

Figure 1. Distribution of active and overlapping Golden and Bonelli's eagle territories in Crete.

3.2. Nest Site Selection & Habitat Suitability

The binary logistic regression model showed that both species build their nests on steep cliffs away from olive groves although the latter effect was marginally non-significant for the Bonelli's eagle (Table 4). Furthermore, the Golden eagle exhibited a wider temperature and altitudinal range away from cultivations while in contrast the Bonelli's eagle that was found breeding on cliff walls in low altitude areas with more vegetation or natural grasslands in their vicinity. The inter-specific nearest neighboring distance was found statistically significant only for the Golden eagle pinpointing an additional important predictor for its nest site selection.

Table 4. Results of the binary logistic regression models investigating the factors that affect nest site selection by the Golden and the Bonelli's eagles on Crete.

Variable	*Aquila chrysaetos*			*Aquila fasciata*		
	Estimate	**z**	***p***	**Estimate**	**z**	***p***
Temperature Annual Range	0.1263	0.24	0.81	1.2467	2.38	0.017
Elevation	-23×10^{-5}	−0.1	0.92	−0.007	−3.49	0.0005
Slope	0.143	3.35	0.0008	0.1756	4.49	7.1×10^{-6}
EVI winter	−12.1123	−1.63	0.1	−12.7893	−2.31	0.021
NNID	29×10^{-5}	2.75	0.0060	82×10^{-6}	1	0.32
Olive groves	−9.2023	−2.92	0.0035	−3.3295	−1.84	0.065
Agriculture	−11.0747	−2.74	0.0062	−2.8417	−1.28	0.2
Natural grassland	2.5209	1.42	0.16	3.5106	2.41	0.02

The MaxEnt model predicted that more suitable nesting habitat for the Golden eagle exists on eastern Crete as well as in the periphery of the major massifs of the island (Figure 2a) whereas that of the Bonelli's is mainly restricted to less mountainous areas and largely on coastal zones (Figure 2b). The prediction accuracy of the model for both species was adequately high (i.e., *A. chrysaetos*: AUC = 0.884 ± 0.02; *A. fasciata*: AUC = 0.880 ± 0.03). Furthermore, the inspection of the overlap between binary presence/absence predictive maps showed that 35% and 60% of suitable nesting habitat is uniquely and potentially available for Golden and Bonelli's eagle nesting, respectively.

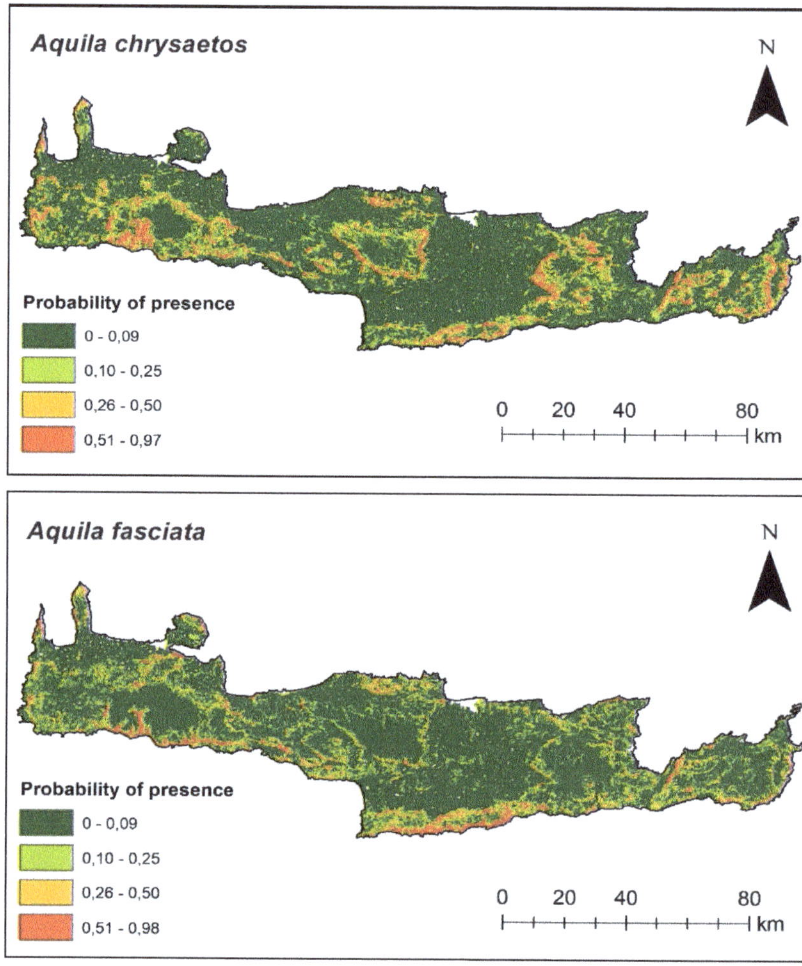

Figure 2. Nesting habitat suitability maps (predicted by a Maximum Entropy approach) and probability of nest site occurrence for Golden and the Bonelli's eagle on Crete.

4. Discussion

In general, the habitat parameters that best discriminate nest site selection between the two species were associated with elevation, temperature, and land cover type, which most probably reflect prey variety and availability [74]. Both species select vertical precipices for breeding which seems self-explanatory for cliff-nesting raptors, though steep cliffs have been regarded as a prerequisite for territory occupancy in the case of the Bonelli's eagle [75]. Univariate analysis and regression models revealed that certain habitat features were also favored by the species in the vicinity of their nest sites such as an intermediate woodland to shrubland for the Golden eagle and natural grasslands for the Bonelli's eagle. These findings should be attributed to the altitudinal differences between the nesting territories of the two species as well as to their main prey items. The Golden eagle nests in mid- and high-altitude areas where bushy and herbaceous vegetation with occasional scattered trees and bare land prevail and its diet consists mainly of brown hares (*Lepus europeus*), chukar partridges (*Alectoris chukar*), and wood doves (*Columba palumbus*), as well as livestock and carrion [18,20,21]. On the contrary, the Bonelli's eagle nests at lower altitude and closer to the coast, while being smaller in size and nimbler, may also hunt Rock doves (*Columba livia*), common pigeons, gulls (*Larus* spp.), corvids, and even passerines (e.g., *Turdus* spp.) [20]. On the other hand, both species avoid olive groves most likely due to the scarcity of available prey in the understory and the lack of suitable cliffs for nesting or uplift soaring [55,56,76,77]. In general, farmland is suboptimum habitat for both species [2,49,57,78,79] although rural areas extensively exploited has been reported to be beneficial for both [59,80,81]. This was also shown in the present study as the Golden eagle selected transitional zones of woodland to shrubland; a habitat with open spaces that can provide ample feeding opportunities to breeding individuals (e.g., livestock carrion, medium-sized mammals). In the same context, natural grasslands that were selected by the Bonelli's eagle are habitats of low productivity under moderate human use and encompass rocky areas on steep slopes and patches of natural vegetation that favors breeding and foraging. Similarly, the olive groves selected by the Bonelli's eagles in overlapping territories are trees cultivated in traditional terraces with patches of natural scrubland in their margins and differ greatly from the monoculture of the plains.

Relevant discrepancies between the two species were also detected for the bioclimatic variables (e.g., temperature, precipitation) prevailing in their nesting habitat. For instance ranges of extreme temperature conditions were more influential on the breeding distribution of the Bonelli's eagle than its congeneric Golden eagle and most probably were associated with altitudinal preferences [26,82]. In particular the ecological niche separation by temperature might be adjusted by the altitude and has been suggested that reflects the species biogeographical background namely Holarctic for the Golden eagle and Indoafrican for the Bonelli's eagle [83–85].This fact was also advocated in the species overlapping territories at mid-altitude areas where no significant differences were noticed in any of the bioclimatic variables examined.

Regarding human pressure, the nests of Bonelli's eagles were located closer to human settlements presumably due to their breeding in lower altitude or their tolerance to a certain degree of human presence and the relatively high availability of their favorable prey (e.g., rabbits, pigeons) in altered environments [14,24,57,85,86]. In contrast, the location of Golden eagle nests is in more remote inland areas, although in line with previous research [2,85,87–90], this should be regarded as a byproduct of direct persecution. The Golden eagle is a top predator on Crete and has been heavily persecuted by stock-breeders because it preys upon newborn lambs. High human-induced mortality is depicted locally in the number of territories near inhabited areas which remain abandoned or occupied by single adult birds or mixed pairs of adult and pre-adult individuals [21,91]. In contrast, this is not the case for the Bonelli's eagle. This species occupies territories in the vicinity of inhabited areas which are located in narrow inaccessible gorges and coastal cliff faces where eagle-man encounters are rare or impeded by the species cryptic character.

In conclusion topographic variables were demonstrated to be most powerful for niche separation between the study species and most significant in predicting suitable nesting habitat on the island. Respective modeling identified a negative correlation of Bonelli's eagle with altitude revealing its lowland and coastal character in contrast to the Golden eagle who proved to be a mountainous and continental species. At the same time, the inclusion of other climatic disturbance or land use variables did not improve the predictive power of the model. However, a number of additional influential factors, i.e., prey diversity, density of power lines, or human infrastructures that were not incorporated in the present study could explain distributional gaps and enhance the interpretability of any possible synergies of human activities and inter-specific competition [57,91–95].

The ecological niche separation facilitated by altitude and consequently temperature has been suggested by other studies to be related to inter-specific competition [85,96] or even the local decline of the Bonelli's eagle [84,97]. In the present study, the two eagle species occupied significantly differential ecological niches selecting distinct nesting habitats at an altitudinal gradient. However, this fact does not inevitably imply an attempt to reduce inter-specific interactions. The PAT and niche similarity analysis did not exclude spatial overlap rather than highlighted differential ecological requirements (e.g., territory size, prey variety, food availability) and close coexistence in 44% of the areas highly used by the Bonelli's eagle in overlapping territories. Besides the inter-specific nearest neighboring distance was found statistically significant only for the Golden eagle, suggesting that this is the less tolerant species in its co-existence with the Bonelli's eagle rather than the other way around. Territorial defense and aggression against intruders have been regularly observed during fieldwork and this was always directed from Bonelli's eagles against Golden eagles. Besides during the study period, no Golden eagle colonized an abandoned Bonelli's eagle territory although floating non-territorial birds of the former regularly visited vacant nesting cliffs of the latter. This fact could be explained by the low persistence of Golden eagles close to human activities most likely due to direct persecution.

In a conservation perspective, it has been supported that management actions, beneficial for both species, should focus on mountain habitats with intermediate temperature, low human density and high prey diversity [84]. In the current case, this is especially so in rural areas where the landscape is dominated by mixed and natural Mediterranean vegetation such as shrubland and semi-natural grasslands or extensive olive-tree farming. In addition, more research is needed on the species home range use and habitat heterogeneity which could favor their coexistence in overlapping or neighboring territories [29,98,99]. Until then, conservation planning should be a priority in an effort to eliminate mortality factors related to human infrastructure (e.g., colliding to power lines or wind energy facilities) or disturbance from construction and operation activities which may lead to breeding failures or the permanent loss of nesting territories.

Author Contributions: Conceptualization, S.M.X.; methodology, S.M.X., M.S., E.T. and E.G.; software, S.M.X., E.T. and M.S.; validation, E.G.; formal analysis, S.M.X. and M.S.; investigation, S.M.X., A.K. and K.D.; data curation, S.M.X., A.K. and K.D.; writing—original draft preparation, M.S. and E.T.; writing—review and editing, S.M.X., M.S., E.T. and E.G.; supervision, S.M.X.; project administration, S.M.X. All authors have read and agreed to the published version of the manuscript.

Funding: This research was funded by LIFE17 NAT/GR/000514, the Green Fund of the Hellenic Ministry of Environment and Energy and the A.G. Leventis Foundation (Project KA 10147 of the Special Account of Research Funds of University of Crete).

Institutional Review Board Statement: Not applicable.

Data Availability Statement: Not applicable.

Conflicts of Interest: The authors declare no conflict of interest. The funders had no role in the design of the study; in the collection, analyses, or interpretation of data; in the writing of the manuscript; or in the decision to publish the results.

References

1. Newton, I. *Population Ecology of Raptors*; T & AD Poyser: Berkhamsted, UK, 1979.
2. Watson, J. The Golden Eagle, 2nd ed.T & AD Poyser: London, UK, 2010; ISBN 1408134543.
3. Carrete, M.; Sánchez-Zapata, J.A.; Calvo, J.F.J.F.; Lande, R. Demography and Habitat Availability in Territorial Occupancy of Two Competing Species. *Oikos* **2005**, *108*, 125–136. [CrossRef]
4. Sergio, F.; Pedrini, P.; Rizzolli, F.; Marchesi, L. Adaptive Range Selection by Golden Eagles in a Changing Landscape: A Multiple Modelling Approach. *Biol. Conserv.* **2006**, *133*, 32–41. [CrossRef]
5. Martínez, J.A.; Calvo, J.F.; Martínez, J.E.; Zuberogoitia, I.; Zabala, J.; Redpath, S.M. Breeding Performance, Age Effects and Territory Occupancy in a Bonelli's Eagle Hieraaetus Fasciatus Population. *Ibis* **2008**, *150*, 223–233. [CrossRef]
6. Bourski, O.V.; Forstmeier, W. Does Interspecific Competition Affect Territorial Distribution of Birds? A Long-Term Study on Siberian Phylloscopus Warblers. *Oikos* **2000**, *88*, 341–350. [CrossRef]
7. Preston, K.L.; Rotenberry, J.T.; Redak, R.A.; Allen, M.F. Habitat Shifts of Endangered Species under Altered Climate Conditions: Importance of Biotic Interactions. *Glob. Chang. Biol.* **2008**, *14*, 2501–2515. [CrossRef]
8. Rodríguez, B.; Rodríguez, A.; Siverio, F.; Siverio, M. Factors Affecting the Spatial Distribution and Breeding Habitat of an Insular Cliff-Nesting Raptor Community. *Curr. Zool.* **2018**, *64*, 173–181. [CrossRef]
9. Carrete, M.; Sánchez-Zapata, J.A.; Benítez, J.R.; Lobón, M.; Donázar, J.A. Large Scale Risk-Assessment of Wind-Farms on Population Viability of a Globally Endangered Long-Lived Raptor. *Biol. Conserv.* **2009**, *142*, 2954–2961. [CrossRef]
10. Cramp, S.; Simmons, K.E.L. *The Birds of the Western Palearctic*; Oxford University Press: Oxford, UK, 1980; ISBN 978-0-19-857506-1.
11. Del Hoyo, J.; Elliott, A.; Sargatal, J. *Handbook of the Birds of the World. Volume 2: New World Vultures to Guinea Fowl*; Lynx Edicions: Barcelona, Spain, 1994.

12. Garza, V.; Arroyo, B. Situación Del Águila Perdicera (Hieraetus Fasciatus) En España. In *Biología y conservación de las Rapaces Mediterráneas*; Muntaner, J., Mayol, J., Eds.; SEO/Birdlife: Madrid, Spain, 1996; pp. 219–230.
13. Real, J.; Mañosa, S. Demography and Conservation of Western European Bonelli's Eagle Hieraaetus Fasciatus Populations. *Biol. Conserv.* **1997**, *79*, 59–66. [CrossRef]
14. Carrete, M.; Sánchez-Zapata, J.A.; Martínez, J.E.; Sánchez, M.Á.; Calvo, J.F. Factors Influencing the Decline of a Bonelli's Eagle Hieraaetus Fasciatus Population in Southeastern Spain: Demography, Habitat or Competition? *Biodivers. Conserv.* **2002**, *11*, 975–985. [CrossRef]
15. Di Vittorio, M.; Sará, M.; López-López, P. Habitat Preferences of Bonelli's Eagles Aquila Fasciata in Sicily. *Bird Study* **2012**, *59*, 207–217. [CrossRef]
16. Arroyo, B.; Ferreiro, E.; Garza, V. *El Águila Perdicera (Hieraaetus Fasciatus) En España: Censo, Reproducción y Conservación*; Instituto Nacional para la Conservación de la Naturaleza: Madrid, Spain, 1995; ISBN 8480140143.
17. Muñoz, A.R.; Real, R. Distribution of Bonelli's Eagle Aquila Fasciata in Southern Spain: Scale May Matter. *Acta Ornithol.* **2013**, *48*, 93–101. [CrossRef]
18. Handrinos, G.; Akriotis, T. *The Birds of Greece*; C. Helm: London, UK, 1997.
19. BirdLife International Species Factsheet: Aquila Chrysaetos. Available online: http://www.birdlife.org (accessed on 24 August 2021).
20. Legakis, A.; Maragou, P. *The Red Data Book of Threatened Animal Species of Greece*; Greek Zoological Society: Athens, Greece, 2009; ISBN 9789608529885.
21. Xirouchakis, S. The Golden Eagle Aquila Chrysaetos in Crete. Distribution, Population Status and Conservation Problems. *Avocetta* **2001**, *25*, 275–281.
22. Bourdakis, S.; Xirouchakis, S.M. The Bonelli's Eagle (Hieraaetus Fasciatus) in Greece. In *The Bonelli's Eagle: Ecology, Behaviour and Conservation*; Hernadez, V., Ed.; Tundra Ediciones: Madrid, Spain, 2009.
23. Panuccio, M.; Agostini, N.; Nelli, L.; Andreou, G.; Xirouchachis, S. Factors Shaping Distribution and Abundance of Raptors Wintering in Two Large Mediterranean Islands. *Community Ecol.* **2019**, *20*, 93–103. [CrossRef]
24. Gil Sanchez, J.; Molino Garrido, F.; Valenzuela Serrano, G. Seleccion de Habitat de Nidificacion Por El Aguila Perdicera (Hieraaetus Fasciatus) En Granada (SE de Espana). *Ardeola* **1996**, *43*, 189–197.
25. Manly, J.J.; Jacobs, D.M.; Touradji, P.; Small, S.A.; Stern, Y. Reading Level Attenuates Differences in Neuropsychological Test Performance between African American and White Elders. *J. Int. Neuropsychol. Soc.* **2002**, *8*, 341–348. [CrossRef]
26. Ontiveros, D.; Pleguezuelos, J.M. Influence of Climate on Bonelli's Eagle's (Hieraaetus Fasciatus V. 1822) Breeding Success through the Western Mediterranean. *J. Biogeogr.* **2003**, *30*, 755–760. [CrossRef]
27. Pianka, E. Competition and Niche Theory. In *Theoretical Ecology Principles and applications*; May, R.M., Ed.; Blackwell: London, UK, 1981; pp. 167–196. ISBN 0878935150.
28. Schoener, T.W. Resource Partitioning in Ecological Communities. *Science* **1974**, *185*, 27–39. [CrossRef]
29. Chesson, P. Mechanisms of Maintenance of Species Diversity. *Annu. Rev. Ecol. Syst.* **2000**, *31*, 343–366. [CrossRef]
30. Donazar, J.A.; Blanco, G.; Hiraldo, F.; Soto-Largo, E.; Oria, J. Effects of Forestry and Other Land-Use Practices on the Conservation of Cinereous Vultures. *Ecol. Appl.* **2002**, *12*, 1445. [CrossRef]
31. Thibault, J.-C.; Patrimonio, O.; Torre, J. Does the Diurnal Raptor Community of Corsica (Western Mediterranean) Show Insular Characteristics? *J. Biogeogr.* **1992**, *19*, 363. [CrossRef]
32. Rackham, O.; Moody, J. *The Making of the Cretan Landscape*; Manchester University Press: Manchester, UK, 1996; ISBN 071903647X.
33. Penna, P. The Climate of Crete, Unpubl. Ph.D. Thesis (In Greek.), Aristotelian University, Salonica, Greece, 1977.
34. Grove, A.T.; Jennifer, M.; Rackham, O.; Grove, A.T.; Moody, J.; Rackham, O. *Crete and the South Aegean Islands: Effects of Changing Climate on the Environment*; Unpubl. report; Robinson Coll. Univ: Cambridge, UK, 1991.
35. Fuller, M.R.; Mosher, J.A. Raptor Survey Techniques. In *Raptor Management Techniques Manual*; Pendleton, B.G., Millsap, B.A., Cline, K.W., Bird, D.M., Eds.; National Wildlife Federation: Washington, DC, USA, 1987; pp. 37–65.
36. Andersen, D.E.; Rongstad, O.J.; Mytton, W.R. Home-Range Changes in Raptors Exposed to Increased Human Activity Levels in Southeastern Colorado. *Wildl. Soc. Bull.* **1990**, *18*, 134–142.
37. Beaver, D.L.; Roth, J.J. Winter Survey of Raptors with Notes on Avian Scavengers in Northwestern Colorado. *Gt. Basin Nat.* **1997**, *57*, 184–186.
38. Williams, C.K.; Applegate, R.D.; Lutz, R.S.; Rusch, D.H. A Comparison of Raptor Densities and Habitat Use in Kansas Cropland and Rangeland Ecosystems. *October* **2000**, *34*, 203–209.
39. Harmata, A.R. What Is the Function of Undulating Flight Display in Golden Eagles? *Raptor Res.* **1982**, *16*, 103–109.
40. Huete, A.; Didan, K.; Miura, T.; Rodriguez, E.P.; Gao, X.; Ferreira, L.G. Overview of the Radiometric and Biophysical Performance of the MODIS Vegetation Indices. *Remote Sens. Environ.* **2002**, *83*, 195–213. [CrossRef]
41. Busetto, L.; Ranghetti, L. MODIStsp: An R Package for Automatic Preprocessing of MODIS Land Products Time Series. *Comput. Geosci.* **2016**, *97*, 40–48. [CrossRef]
42. R Core Team. R: A Language and Environment for Statistical Computing. R Foundation for Statistical Computing. Available online: http://www.R-project.org/ (accessed on 20 July 2021).
43. European Union Copernicus Land Monitoring Service. Eur. Environ. Agency 1990. Available online: https://land.copernicus.eu/pan-european/corine-land-cover (accessed on 12 August 2021).

44. European Union Copernicus Land Monitoring Service. Eur. Environ. Agency. 2012. Available online: https://land.copernicus.eu/pan-european/corine-land-cover (accessed on 12 August 2021).
45. Beecham, J.; Kochert, M.N. Breeding Biology of the Golden in Southwestern Idaho. *Wilson Bull.* **1975**, *85*, 506–513.
46. Phillips, R.L.; McEneaney, T.P.; Beske, A.E. Population Densities of Breeding Golden Eagles in Wyoming. *Wildl. Soc. Bull.* **1984**, *12*, 269–273.
47. Collopy, M.W.M.; Edwards Jr, T.C.; Edwards, T.J. Territory Size, Activity Budget, and Role of Undulating Flight in Nesting Golden Eagles. *J. Field Ornithol.* **1989**, *60*, 43–51.
48. Arroyo, A.C.; Solís, C.D.L.C. Individual and Seasonal Variability in Territory Size of Three Adult Bonelli's Eagles Aquila Fasciata in Extremadura (Sw Spain). *Ardeola* **2011**, *58*, 165–173. [CrossRef]
49. Katzner, T.; Smith, B.W.; Miller, T.A.; Brandes, D.; Cooper, J.; Brauning, D.; Farmer, C.; Harding, S.; Kramar, D.E.; Koppie, C.; et al. America's Eastern Golden Eagle (Aquila Chrysaetos) Population. *Auk* **2012**, *129*, 168–176.
50. Rosenberg, D.K.; McKelvey, K.S. Estimation of Habitat Selection for Central-Place Foraging Animals. *J. Wildl. Manag.* **1999**, *63*, 1028–1038. [CrossRef]
51. McLeod, D.R.A.; Whitfield, D.P.; McGrady, M.J. Improving Prediction of Golden Eagle (Aquila Chrysaetos) Ranging in Western Scotland Using GIS and Terrain Modeling. *J. Raptor Res.* **2002**, *36*, 70–77.
52. Bosch, R.; Real, J.; Tint, A.; Zozaya, L.E.; Castell, C. Home-ranges and Patterns of Spatial Use in Territorial Bonelli's Eagles Aquila Fasciata. *Ibis* **2010**, *152*, 105–117. [CrossRef]
53. Zar, J. *Biostatistical Analysis*, 5th ed.; Prentice-Hall/Pearson, Ed.; Scientific Research Publishing: Upper Saddle River, NJ, USA, 2010; Volume xiii.
54. Mcleod, D.R.A.; Whitfield, D.P.; Fielding, A.H.; Haworth, P.F.; Mcgrady, M.J. Predicting Home Range Use by Golden Eagles Aquila Chrysaetos in Western Scotland. *Avian Sci.* **2002**, *2*, 183–198.
55. Mcleod, D.R.A.; Whitfield, D.P. GIS for Golden Eagle Management. Scottish Natural Heritage: Edinburgh, UK, 1999.
56. McGrady, M.J.; Grant, J.R.; Bainbridge, I.P.; McLeod, D.R.A. A Model of Golden Eagle (Aquila Chrysaetos) Ranging Behavior. *J. Raptor Res.* **2002**, *36*, 62–69.
57. López-López, P.; García-Ripollś, C.; Aguilar, J.M.; Garcia-López, F.; Verdejo, J. Modelling Breeding Habitat Preferences of Bonelli's Eagle (Hieraaetus Fasciatus) in Relation to Topography, Disturbance, Climate and Land Use at Different Spatial Scales. *J. Ornithol.* **2006**, *147*, 97–106. [CrossRef]
58. Arroyo, B.; Ferreiro, E.; Garza, V. *El Aguila Real (Aquila Chrysaetos) En Espana. Censo, Distribucion y Conservacion*; ICONA: Madrid, Spain, 1990.
59. Di Vittorio, M.; López-López, P. Spatial Distribution and Breeding Performance of Golden Eagles Aquila Chrysaetos in Sicily: Implications for Conservation. *Acta Ornithol.* **2014**, *49*, 33–45. [CrossRef]
60. Sanz, A.S.; Ínguez, E.M.; Anadón, J.D.A.; Ernández, V.J.H.; Sanz-Aguilar, A.; Mínguez, E.; Anadón, J.D.A.; Hernández, V. Uso Heterogéneo Del Espacio En Tres Territorios de Reproducción Del Aguila-Azor Perdicera (Hieraaetus Fasciatus). *Ardeola Rev. ibérica Ornitol.* **2005**, *52*, 347–350.
61. Warren, C. *Managing Scotland's Environment*; Edinburgh University Press: Edinburgh, UK, 2022; ISBN 9780748630639.
62. Schoener, T.W. Sizes of Feeding Territories among Birds. *Ecology* **1968**, *49*, 123–141. [CrossRef]
63. Warren, D.; Glor, R.; Turelli, M. ENMTools: A Toolbox for Comparative Studies of Environmental Niche Models. *Ecography* **2010**, *33*, 607–611. [CrossRef]
64. Warren, D.L.; Matzke, N.J.; Cardillo, M.; Baumgartner, J.B.; Beaumont, L.J.; Turelli, M.; Glor, R.E.; Huron, N.A.; Simões, M.; Iglesias, T.L.; et al. ENMTools 1.0: An R Package for Comparative Ecological Biogeography. *Ecography* **2021**, *44*, 504–511. [CrossRef]
65. Aiello-Lammens, M.E.; Boria, R.A.; Radosavljevic, A.; Vilela, B.; Anderson, R.P. SpThin: An R Package for Spatial Thinning of Species Occurrence Records for Use in Ecological Niche Models. *Ecography* **2015**, *38*, 541–545. [CrossRef]
66. Montgomery, D.C.; Peck, E.A. *Introduction to Linear Regression Analysis*, 2nd ed.; Wiley: Hoboken, NJ, USA, 1992.
67. Kovács, G.; Spens, K.M. Abductive Reasoning in Logistics Research. *Int. J. Phys. Distrib. Logist. Manag.* **2005**, *35*, 132–144. [CrossRef]
68. Quinn, G.P.; Keough, M.J. *Experimental Design and Data Analysis for Biologists*; Cambridge University Press: New York, NY, USA, 2002.
69. Boyce, M.S.; Vernier, P.R.; Nielsen, S.E.; Schmiegelow, F.K.A. Evaluating Resource Selection Functions. *Ecol. Modell.* **2002**, *157*, 281–300. [CrossRef]
70. Phillips, S.; Anderson, R.; Dudík, M.; Schapire, R.; Blair, M. Opening the Black Box: An Open-Source Release of Maxent. *Ecography* **2017**, *40*, 887–893. [CrossRef]
71. Hand, D. Measuring Classifier Performance: A Coherent Alternative to the Area Under the ROC Curve. *Mach. Learn.* **2009**, *77*, 103–123. [CrossRef]
72. Melo, F. Receiver Operating Characteristic (ROC) Curve. In *Encyclopedia of Systems Biology*; Springer: New York, NY, USA, 2013; pp. 1818–1823.
73. Franklin, J. *Mapping Species Distributions*; Cambridge University Press: Cambridge, UK, 2010; ISBN 9780511810602.
74. López-López, P.; Urios, V. Use of Digital Trail Cameras to Study Bonelli's Eagle Diet during the Nestling Period. *Ital. J. Zool.* **2010**, *77*, 289–295. [CrossRef]

75. Sánchez-Zapata, J.A. Recovery Plan for Bonelli's Eagle in Murcia, Spain. *Ambient. S.L.-Consejería Agric,* Agua y Medio Ambient: Murcia, Spain, 1996.
76. Singh, N.J.; Moss, E.; Hipkiss, T.; Ecke, F.; Dettki, H.; Sandström, P.; Bloom, P.; Kidd, J.; Thomas, S.; Hörnfeldt, B. Habitat Selection by Adult Golden Eagles Aquila Chrysaetos during the Breeding Season and Implications for Wind Farm Establishment. *Bird Study* **2016**, *63*, 233–240. [CrossRef]
77. Fielding, A.H.; Haworth, P.F.; Anderson, D.; Benn, S.; Dennis, R.; Weston, E.; Whitfield, D.P. A Simple Topographical Model to Predict Golden Eagle Aquila Chrysaetos Space Use during Dispersal. *Ibis* **2020**, *162*, 400–415. [CrossRef]
78. Watson, J. Golden Eagle Aquila Chrysaetos Breeding Success and Afforestation in Argyll. *Bird Study* **1992**, *39*, 203–206. [CrossRef]
79. Carrete, M.; Sánchez-Zapata, J.A.; Calvo, J.F. Breeding Densities and Habitat Attributes of Golden Eagles in Southeastern Spain. *J. Raptor Res.* **2000**, *34*, 48–52.
80. Tapia, L.; Domínguez, J.; Rodríguez, L. *Modelling Habitat Use and Distribution of Golden Eagles Aquila Chrysaetos in a Low-Density Area of the Iberian Peninsula BT—Biodiversity and Conservation in Europe*; Hawksworth, D.L., Bull, A.T., Eds.; Springer: Dordrecht, The Netherlands, 2008; pp. 317–332. ISBN 978-1-4020-6865-2.
81. Martínez-Miranzo, B.; Banda, I.E.; Gardiazábal, A.; Ferreiro, E.; Seoane, J.; Aguirre, J.I. Long-term Monitoring Program Reveals a Mismatch between Spatial Distribution and Reproductive Success in an Endangered Raptor Species in the Mediterranean Area. *Ecol. Res.* **2022**, *37*, 421–431. [CrossRef]
82. Muñoz, A.R.; Real, R.; Barbosa, A.M.; Vargas, J.M. Modelling the Distribution of Bonelli's Eagle in Spain: Implications for Conservation Planning. *Divers. Distrib.* **2005**, *11*, 477–486. [CrossRef]
83. Ferguson-Lees, J.; Christie, D. *Raptors of the World*; Hought. Mifflin Company: New York, NY, USA, 2001. [CrossRef]
84. Moreno-Rueda, G.; Pizarro, M.; Ontiveros, D.; Pleguezuelos, J.M. The Coexistence of the Eagles Aquila Chrysaetos and Hieraaetus Fasciatus Increases with Low Human Population Density, Intermediate Temperature, and High Prey Diversity. *Ann. Zool. Fennici* **2009**, *46*, 283–290. [CrossRef]
85. López-López, P.; García-Ripollés, C. Patrón de Distribución Del Águila Real (Aquila Chrysaetos) y Del Águila-Azor Perdicera (Hieraaetus Fasciatus) En La Provincia de Castellón. *Ardeola* **2004**, *51*, 275–283.
86. López-López, P.; García-Ripollés, C.; Soutullo, Á.; Cadahía, L.; Urios, V. Identifying Potentially Suitable Nesting Habitat for Golden Eagles Applied to "important Bird Areas" Design. *Anim. Conserv.* **2007**, *10*, 208–218. [CrossRef]
87. Kaisanlahti-Jokimäki, M.L.; Jokimäki, J.; Huhta, E.; Ukkola, M.; Helle, P.; Ollila, T. Territory Occupancy and Breeding Success of the Golden Eagle (Aquila Chrysaetos) around Tourist Destinations in Northern Finland. *Ornis Fenn.* **2008**, *85*, 2–12.
88. Martin, J.; McIntyre, C.L.; Hines, J.E.; Nichols, J.D.; Schmutz, J.A.; MacCluskie, M.C. Dynamic Multistate Site Occupancy Models to Evaluate Hypotheses Relevant to Conservation of Golden Eagles in Denali National Park, Alaska. *Biol. Conserv.* **2009**, *142*, 2726–2731. [CrossRef]
89. Lado, L.R.; Tapia, L. Suitable Breeding Habitat for Golden Eagle (Aquila Chrysaëtos) in a Border of Distribution Area in Northwestern Spain: Advantages of Using Remote Sensing Information vs. Land Use Maps. *Vie Milieu* **2012**, *62*, 77–85.
90. Djorgova, N.; Ragyov, D.; Biserkov, V.; Biserkov, J.; Nikolov, B.P. Habitat Preferences of Diurnal Raptors in Relation to Human Access to Their Breeding Territories in the Balkan Mountain Range, Bulgaria. *Avian Res.* **2021**, *12*, 29. [CrossRef]
91. Xirouchakis, S. Causes of Raptor Mortality in Crete. In *Raptors Worldw*; World Working Group on Birds of Prey/MME: Berlin, Germany, 2004; pp. 849–860.
92. Mañosa, S.; Real, J.; Codina, J. Selection of Settlement Areas by Juvenile Bonelli's Eagle in Catalonia. *J. Raptor Res.* **1998**, *32*, 208–214.
93. Moleón, M.; Bautista, J.; Sánchez-Zapata, J.A.; Gil-Sánchez, J.M. Diet of Non-breeding Bonelli's Eagles Hieraaetus Fasciatus at Settlement Areas of Southern Spain. *Bird Study* **2009**, *56*, 142–146. [CrossRef]
94. Rollan, A.; Real, J.; Bosch, R.; Tinto, A.; Hernandez-Matias, A. Modelling the Risk of Collision with Power Lines in Bonelli's Eagle Hieraaetus Fasciatus and Its Conservation Implications. *Bird Conserv. Int.* **2010**, *20*, 279–294. [CrossRef]
95. Di Marzio, A.; Lambertucci, S.A.; García-Fernández, A.J.; Martínez-López, E. Temporal Changes in Metal Concentrations in Andean Condor Feathers: A Potential Influence of Volcanic Activity. *Environ. Sci. Pollut. Res.* **2020**, *27*, 25600–25611. [CrossRef]
96. López-López, P.; Ferrer, M.; Madero, A.; Casado, E.; McGrady, M. Solving Man-Induced Large-Scale Conservation Problems: The Spanish Imperial Eagle and Power Lines. *PLoS ONE* **2011**, *6*, e17196. [CrossRef]
97. Gil-Sánchez, J.M. Parametros Reproductivos y de Alimentacion Del Aguila Real (Aquila Chrysaetos) y Del Aguila Perdicera (Hieraaetus Fasciatus) En La Provincia de Granada. *Ardeola* **1994**, *12*, 47–52.
98. Real, J. Águila-Azor Perdicera, Hieraaetus Fasciatus. In *Libro Rojo de Las Aves de España*; Madroño, A., Moreno-Rueda, G., González, C., Atienza, J.., Eds.; Dirección General para la Biodiversidad-SEO/Birdlife: Madrid, Spain, 2004; pp. 154–157.
99. Marzluff, J.M.; Vekasy, T.K.M.S.; Schueck, L.S.; Zarriello, T.J.; Knick, S.T.; Vekasy, M.S.; Schueck, L.S.; Zarriello, T.J. Spatial Use and Habitat Selection of Golden Eagles in Southwestern Idaho. *Auk* **1997**, *114*, 673–687. [CrossRef]

diversity

MDPI

Article

Home-Range Size and Space Use of Territorial Bonelli's Eagles (*Aquila fasciata*) Tracked by High-Resolution GPS/GSM Telemetry

Sara Morollón [1], Vicente Urios [1] and Pascual López-López [2,]*

1 Grupo de Investigación Zoología de Vertebrados, Campus San Vicente del Raspeig, Universidad de Alicante, Edificio Ciencias III, 03080 Alicante, Spain
2 Movement Ecology Lab., Cavanilles Institute of Biodiversity and Evolutionary Biology, University of Valencia. C/Catedrático José Beltrán 2, Paterna, 46980 Valencia, Spain
* Correspondence: pascual.lopez@uv.es

Abstract: High-resolution GPS/GSM dataloggers provide spatial information of the highest quality, which outperform previous tracking methods, such as Argos telemetry or conventional VHF ground-tracking. As a result, this has improved our knowledge of home-range behavior and spatial ecology of many species, including large raptors. In this paper, we use high-resolution GPS/GSM dataloggers to assess the home-range size and the role of sex, season (breeding or non-breeding season), and breeding status (reproductive or non-reproductive individuals) on the space use of Bonelli's eagle (*Aquila fasciata*). To this end, 51 territorial individuals (25 females and 26 males) were equipped with GPS/GSM transmitters and were tracked over 7 years (2015–2021) in eastern Spain. Overall, we recorded 4,791,080 fixes that were analyzed through kernel density methods (50%, 75%, and 95% fixed kernels). The average individual home-range size according to the 95%, 75%, and 50% kernels was 54.84 ± 20.78 km^2, 24.30 ± 10.18 km^2, and 11.17 ± 4.90 km^2, respectively. Overall, the home-range size of individuals occupying the same territory was similar, mainly due to the cooperative hunting behavior exhibited by the species. We did not find interannual differences in the home-range size (95% fixed kernel) of the majority of individuals, showing a strong territorial fidelity of the breeding pairs. In general, females' home-range size was slightly smaller than males' size due to the decrease in activity in the breeding season as a result of laying, incubation, and chick attendance at nests. No seasonal variation in the 95% kernel was found, but it was found in the 75% and 50% kernels. In regard to the breeding status, higher home-range size was recorded in the non-reproductive individuals. Moreover, we found a low neighbor overlap among the territories (4.18% ± 3.06%), which evidences a high level of intraspecific competition in the Bonelli's eagle. Finally, this study highlights the advantages of the use of accurate telemetry information to improve our understanding of the spatial ecology of the endangered Bonelli's eagle, which ultimately will serve to better inform management actions for its conservation.

Keywords: animal behavior; birds of prey; datalogger; kernel density estimators; movement ecology; Spain

Citation: Morollón, S.; Urios, V.; López-López, P. Home-Range Size and Space Use of Territorial Bonelli's Eagles (*Aquila fasciata*) Tracked by High-Resolution GPS/GSM Telemetry. *Diversity* **2022**, *14*, 1082. https://doi.org/10.3390/d14121082

Academic Editors: Dimitar A. Demerdzhiev, Dimitar Dimitrov and José L. Tella

Received: 8 October 2022
Accepted: 5 December 2022
Published: 8 December 2022

Publisher's Note: MDPI stays neutral with regard to jurisdictional claims in published maps and institutional affiliations.

1. Introduction

The estimation of the home-range size and ranging behavior of predators is of great interest for their conservation, particularly endangered species. This is the case of the Bonelli's eagle (*Aquila fasciata*), a long-lived medium-sized raptor classified as "Near Threatened" (NT) in Europe [1] and "Vulnerable" (VU) in Spain (RD 139/2011). This raptor has a large worldwide distribution, from the Iberian peninsula to Southeast Asia [2]. In Spain, this species usually occupies Mediterranean habitats with evergreen forests and an abrupt orography where it found adequate places for nesting in cliffs and occasionally on trees. The Bonelli's eagle shows a strongly territorial behavior, defending its territory in pairs, where both individuals usually remain together [3–6].

Previous research works in regard to the home-range size and ranging behavior of the Bonelli's eagle were mostly based on radio-tracking technology [7] or based on GPS-Argos telemetry [8,9]. Recently, the use of GPS/GSM dataloggers allows for a large number of precise locations per day, thus favoring the advance of our understanding of the species' ecology.

This study provides a detailed assessment of space use and home-range size of the territorial Bonelli's eagle in Mediterranean landscapes. In addition, we further explored the role of sex, seasonal variations, breeding status, and the neighbor competitive pressure on home-range size. This is the first time that this information has been assessed with high-resolution GPS/GSM technology incorporating the highest accuracy. Therefore, this investigation provides new information on the spatial ecology of the species while taking advantage of a larger sample size in terms of individuals and number of locations in comparison to previous works. As a result, the main goals of this study are to (i) describe the home-range size and ranging behavior of territorial Bonelli's eagles; (ii) examine the influence of sex, season (breeding/non-breeding season), and breeding status (reproductive/non-reproductive individuals) in spatial ecology; and (iii) assess the territorial interaction between neighboring individuals.

2. Materials and Methods

2.1. Study Area

The study area is located in eastern Spain, including Albacete, Alicante, Castellón, Cuenca, and Valencia provinces. The area covers approximately 7600 km^2 with an average altitude ranging between the coastline and 1200 m asl. The climate is Mediterranean with an average annual temperature that varies between 17 °C in the coastal areas and 8 °C in the inner mountains. The dominant landscape is composed of Mediterranean scrublands (*Pistacia lentiscus*, *Rosmarinus officinalis*, *Thymus vulgaris*, *Stipa tenacissima*), oak forests (*Quercus ilex*, *Q. faginea*, *Q. suber*), and Mediterranean evergreen forests (*Pinus halepensis*, *P. pinaster*, *P. nigra*).

2.2. Tracking

A total of 51 territorial adult and subadult Bonelli's eagles, 26 males and 25 females, were trapped in 22 different territories (Figure 1) by means of a remotely activated folding net between 2015 and 2021 (see Table S1, in Supplementary Materials). The trap was always under surveillance by the researchers which were hidden nearby. The trap was only activated once it was checked that the target individuals were inside. Both pair members of each territory, male and female, were trapped together in most cases. When a turnover in one of the pair members was detected (usually after the death of the former member), we trapped and marked the new member. All individuals were tagged with GPS/GSM solar energy dataloggers manufactured by e-obs GmbH (Munich, Germany) and Ornitela (Vilnius, Lithuania) using a backpack configuration by means of a Teflon tubular harness, which is designed to ensure that the harness fell off at the end of the tag's life. The weight of the transmitters was 48 and 50 g, respectively, and represented 1.66% to 2.86% (average = 2.25%, SD = 0.38%) of the body mass of eagles, below the 3% threshold established to avoid negative effects on behavior [10,11]. The duty cycle of the transmitters was programmed to record a GPS location at 5-min intervals [4,5], from 1 h before sunrise to 1 h after sunset, year-round. Moreover, transmitters recorded one fix per hour during nighttime. Transmitters' data were retrieved, stored, and managed by means of the Movebank online repository (http://www.movebank.org/ (accessed on 2 July 2022)).

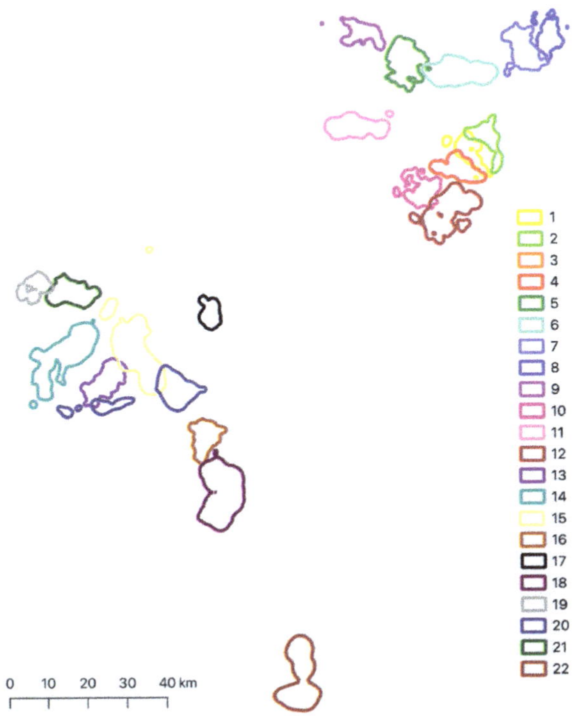

Figure 1. Spatial arrangement of 22 Bonelli's eagle territories in the study area. Due to conservation reasons, the map is hidden to avoid the exact location of each territory.

2.3. Ethical Statement

Trapping and marking activities were authorized and conducted under permissions issued by regional authorities (Generalitat Valenciana and Junta de Comunidades de Castilla-La Mancha, Spain) and all efforts were made to minimize handling time to avoid any suffering to the eagles.

2.4. Home-Range Analysis

We used kernel density estimation (KDE) methods [12] to compute home-range size using the "reproducible home-range" (rhr) R package [13,14]. The KDE methods are widely used to calculate home-range size in different taxonomic groups, such as in fishes (e.g., [15]), reptiles (e.g., [16]), mammals (e.g., [17–22]), and birds (e.g., [4,5,23–28]). We preferably used KDE rather than autocorrelated KDE (AKDE) methods since in our experience, computations based on AKDE methods resulted in an unrealistic larger home-range size (debated in Ref. [29]).

We considered three kernel levels: The 50% kernel (K50% isopleth) which is the nuclear area of the home-range, where the nest is usually located; the 75% kernel (K75% isopleth) which is considered as the intermediate area of active use, and includes the feeding and resting areas; and the 95% kernel (K95% isopleth) which is considered as the total area of the home-range [30,31]). We used the daily individual K50%, K75%, and K95% isopleths computed with the whole tracked period from the following day after tagging to the end day of data transmission (e.g., animal's death, end of transmission), with the knowledge that 15 days is the minimum tracking period to obtain a complete home-range [5].

2.5. Data Modeling

Overall, we recorded 4,791,080 high-precision GPS locations of the 51 adult Bonelli's eagles (26 males and 25 females) that were tracked during the period 2015–2021.

We used generalized linear mixed models (GLMM; [32]) to analyze the variation of the monthly average of daily home-range size within each territory, considering "Territory" and "Individual" as random factors. "Individual" was nested into "Territory" to account for the hierarchical structure (i.e., non-independence) of data [33]. The variation in home-range size was examined in relation to three fixed binary factors: "Sex", "Breeding", and "Season". We considered "Breeding" as the annual breeding status of the pairs, taking into account the years in which each pair did reproduce ("Yes") and those in which they did not ("No"). "Season" was determined using individuals' data combining fieldwork observations and detailed tracking information, and was divided into two different periods: Breeding season and non-breeding season. The "Breeding season" was considered from January to June (since all pairs breed within these months) and the "Non-breeding season" from July to December. The monthly average of daily home-range sizes according to the three different spatial estimators (K95%, K75%, and K50%) were logarithmically transformed and were used as the response variable in the GLMMs. The R package used for the analyses was "lme4" [34].

We used a one-way ANOVA test to analyze the interannual differences in the monthly average of daily home ranges (95% kernel) of the individuals occupying the same territory in different years (i.e., individuals tracked for at least 2 years).

Finally, we calculated the annual territorial overlap between the individuals who occupy simultaneously (i.e., in the same year) neighboring territories. The annual percentage of overlap between neighboring territories was compared using the 95% kernel under the "raster" R package [35].

3. Results

3.1. Individual Home-Range Size

The average K95%, K75%, and K50% were 54.86 ± 20.57 km^2 (range: 22.44–116.11 km^2), 23.91 ± 10.47 km^2 (range: 10.14–56.42 km^2), and 11.17 ± 4.86 km^2 (range: 4.59–27.32 km^2), respectively (Figure S1 and Table S2 in Supplementary Materials).

3.2. Differences in Home-Range Size

Our results showed that home-range size can be explained by the additive effects of "Sex", "Breeding", and "Season" as well as the interactions between "Sex" and "Season" and "Breeding" and "Season". The interaction between "Sex" and "Breeding" was not significant across the three different spatial estimators. The multiple interactions between the three effects were not significant across the three spatial estimators (K95%, K75%, and K50%) (Table 1). Results of $R^2_{conditional}$ corresponding to the effect of combined fixed and random factors were 0.259, 0.282, and 0.295 according to K95%, K75%, and K50%, respectively. The variance explained by random factors (obtained as $R^2_{conditional}$–$R^2_{marginal}$) was 0.217, 0.226, and 0.228 according to K95%, K75%, and K50%, respectively.

In general, taking into account all individuals, the home-range size (K95%, K75%, and K50%) was larger in males than in females ($p < 0.001$ in all cases). Moreover, it was larger during the non-breeding season (only K75% with $p = 0.012$ and K50% with $p = 0.001$) and in the years that the pair did not breed ($p = 0.030$; $p = 0.004$; and $p = 0.001$, respectively) (Table 1).

Taking into account the Season and the Sex, females showed a smaller monthly average daily home-range size than males during the breeding season (Tables 1 and 2; Figure 2). In contrast, there were no differences in home-range size during the non-breeding season (Tables 1 and 2; Figure 2).

Table 1. Generalized linear mixed model (GLMM) results of variation in home-range size using three different spatial estimators. Significant variables are highlighted in bold.

Dep. Variable	Indep. Variable	Estimate	Std. Error	t	d.f	*p*-Value
K95%	(Intercept)	3.817	0.071	53.729	20.727	<0.001
	Sex (Female)	−0.084	0.023	−3.608	24.589	0.001
	Breeding (No)	0.042	0.019	2.179	1234.838	0.030
	Season (Breeding)	−0016	0.017	−0.899	1221.358	0.369
	Sex (Female) × Breeding (No)	0.017	0.018	0.948	707.351	0.343
	Sex (Female) × Season (Breeding)	−0048	0.017	−2.761	1222.184	0.006
	Breeding (No) × Season (Breeding)	0.094	0.018	5.373	1232.930	<0.001
	Sex (Female) × Breeding (No) × Season (Breeding)	0.011	0.017	0.616	1226.767	0.538
K75%	(Intercept)	2.974	0.076	39.063	20.855	<0.001
	Sex (Female)	−0.100	0.025	−3.906	24.274	0.001
	Breeding (No)	0.057	0.020	2.851	1237.418	0.004
	Season (Breeding)	−0.045	0.018	−2.513	1220.620	0.012
	Sex (Female) × Breeding (No)	0.026	0.019	1.349	753.480	0.178
	Sex (Female) × Season (Breeding)	−0.059	0.018	−3.274	1221.557	0.001
	Breeding (No) × Season (Breeding)	0.107	0.018	5.823	1232.456	<0.001
	Sex (Female) × Breeding (No) × Season (Breeding)	0.017	0.018	0.957	1227.514	0.339
K50%	(Intercept)	2.180	0.078	28.047	20.908	<0.001
	Sex (Female)	−0.106	0.027	−4.012	24.016	0.001
	Breeding (No)	0.067	0.020	3.313	1238.865	0.001
	Season (Breeding)	−0.063	0.018	−3.422	1220.088	0.001
	Sex (Female) × Breeding (No)	0.028	0.019	1.467	781.441	0.143
	Sex (Female) × Season (Breeding)	−0.063	0.018	−3.440	1221.091	0.001
	Breeding (No) × Season (Breeding)	0.116	0.019	6.252	1232.181	<0.001
	Sex (Female) × Breeding (No) × Season (Breeding)	0.019	0.018	1.055	1227.970	0.292

Table 2. Monthly average (± standard deviation) daily home-range size (km^2) of the three levels (K95%, K75%, and K50%) taking into account "Season", "Sex", and "Breeding Status".

Season	Sex	Breeding Status	n	K95%	K75%	K50%
Breeding	Females	-	283	48.30 ± 30.60	20.60 ± 14.40	9.21 ± 6.65
Breeding	Males	-	319	62.80 ± 46.70	27.40 ± 22.60	12.40 ± 10.80
Non-Breeding	Females	-	318	56.20 ± 35.30	25.30 ± 17.30	11.70 ± 8.30
Non-Breeding	Males	-	337	57.20 ± 29.80	25.60 ± 13.90	11.80 ± 6.62
Breeding	-	No	270	61.50 ± 35.60	26.90 ± 17.00	12.30 ± 8.13
Breeding	-	Yes	332	51.60 ± 43.70	22.10 ± 21.10	9.77 ± 9.90
Non-Breeding	-	No	288	51.40 ± 32.10	23.20 ± 15.30	10.70 ± 7.28
Non-Breeding	-	Yes	367	60.80 ± 32.40	27.20 ± 15.60	12.50 ± 7.54

Taking into account the Season and the Breeding Status, non-reproductive individuals showed a larger monthly average daily home-range size than reproductive individuals during the breeding season (Tables 1 and 2; Figure 2). During the non-breeding season, the reproductive individuals showed a larger monthly average daily home-range size than non-reproductive individuals (Tables 1 and 2; Figure 2). More detailed comparisons are available in Supplementary Material (Figures S2–S12).

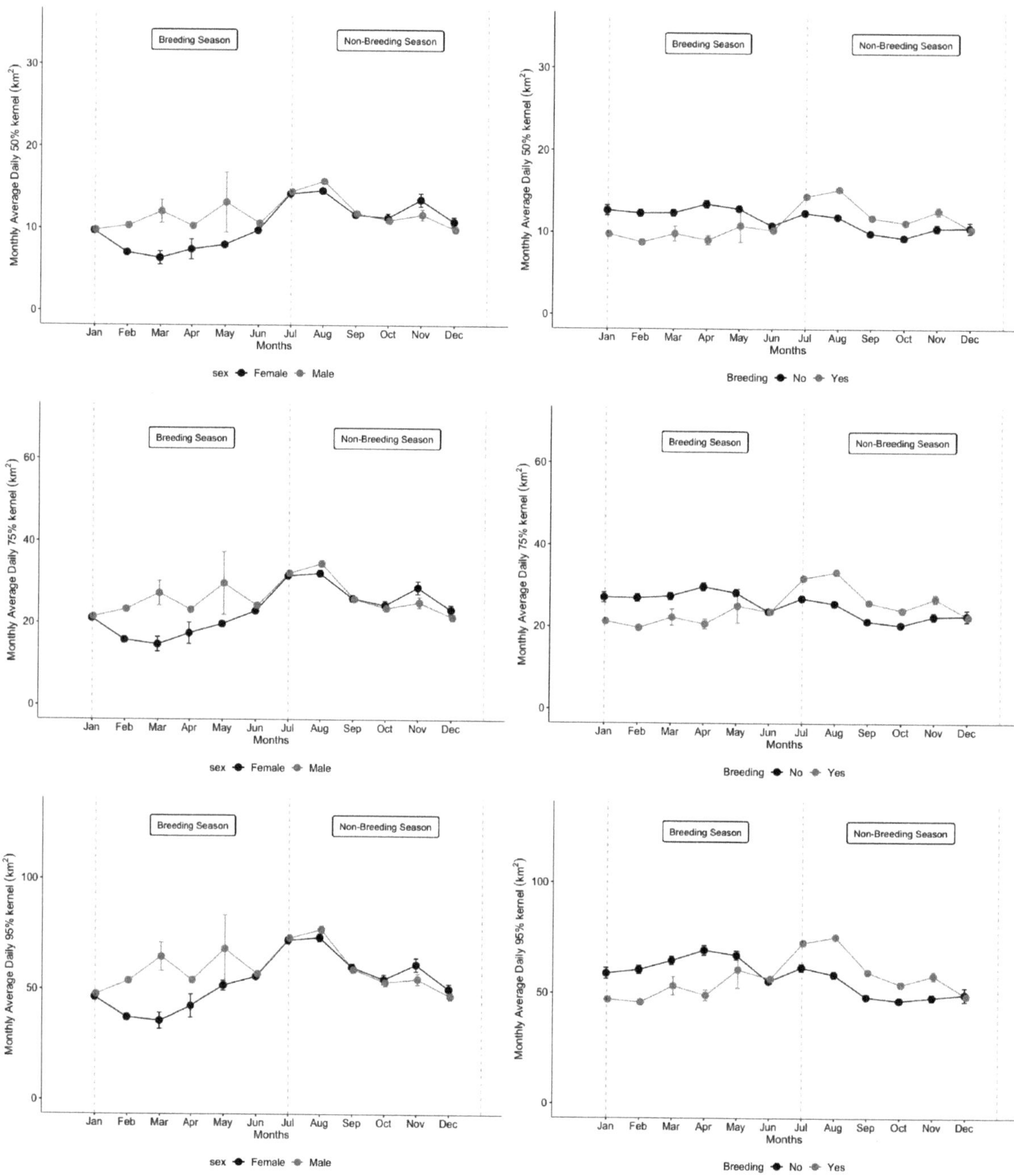

Figure 2. Seasonal differences between sexes (**left**) and breeding status (**right**) in the monthly average of daily home-range size (K95%, K75%, and K50%) over the year. Vertical dashed black lines delimitate the different seasons.

Our results show an inflection point in the monthly home-range size pattern in June and December, both when comparing sexes and breeding status. Females increase their home-range in June, which is similar to males until December, then it decreases again. In regard to the breeding status, breeding individuals increase their home-range in June and decrease in December, while the opposite occurs for non-breeding individuals (Figure 2).

There were no interannual differences in the monthly average of daily home-range size according to the 95% kernel in the majority of the individuals who were tracked for 2 years or more (68.18%, n = 44) (Table S3 in Supplementary Materials).

3.3. Overlapping between Neighboring Territories

The overlap between neighbor home-ranges (K95%) of all individuals (n = 38) showed an average value of 4.18% ± 5.53% (Table 3). Five territories (numbers 9, 11, 14, 17, and 22) were not used in this analysis since they had no neighboring territories occupied in the same year or since there were isolated.

Table 3. Average and standard deviations (SD) of overlap percentages between simultaneous neighboring territories.

Territory 1 (T1)	Territory 2 (T2)	Average Overlap % T1-T2	SD Overlap % T1-T2	Average Overlap % T2-T1	SD Overlap % T2-T1
1	2	3.71	3.76	3.60	3.14
1	3	0.15	NA	0.07	NA
1	4	9.95	1.85	10.71	2.19
1	12	0.49	NA	0.30	NA
2	3	2.67	3.21	1.66	2.30
6	3	5.23	3.86	3.98	3.05
8	7	0.72	0.33	0.45	0.24
10	12	0.72	0.33	1.96	0.85
15	13	1.12	0.52	2.76	1.70
15	20	19.74	12.03	9.30	3.06
16	18	1.09	1.24	0.55	0.61
20	13	4.68	2.43	20.17	19.18
21	19	1.19	0.51	1.70	0.87
Total				4.18	5.53

4. Discussion

Our results show that Bonelli's eagles extend their home-range 54.84 ± 20.78 km^2, which ranges between 22 and 116 km^2 according to the 95% kernel. These results are larger than those described by [36] in Valencia (30.5 km^2; range: 15.82–44.48 km^2) and [7] in Catalonia (36.1 km^2; range: 33.4–110.7 km^2), with both using radio-tracking techniques. In contrast, our results are similar to those reported by [8] in Valencia and Tarragona (44.4 ± 15.4 km^2; range: 31.8–91.9 km^2) and [9] in Aragón, with both using GPS-Argos telemetry (57.25 km^2; range: 23.48–152.24 km^2). The higher the accuracy of GPS locations in comparison to the VHF radio-tracking accounts for the differences observed between the methods [37].

The high variability in the home-range size across territories may be due to different factors, such as geographical differences [38], pressure from neighboring pairs, population density [3,39], interactions with humans [23], and/or different quality of territories [40]. This indicates that there might be territories with high availability of prey that allow pairs to survive for years, while others can be up to three or four times larger.

4.1. Differences in Home-Range Size between Sexes and Seasons

Overall, males showed larger home-range size than females. However, these sex differences were not previously described in the literature [7–9,36]. Sexual differences could be explained by the decrease in the home-range size of the females during the breeding season since they move less due to the fact that they are in charge of the incubation and chicks' attendance [4]. The decrease in this period affects the general size of females' home-range. In contrast, during the breeding period, males are in charge of prospecting the home-range for hunting and delivering the prey to the nest to feed the female and the chicks [4].

In contrast to previous studies, we found no differences in home-range size between the breeding and non-breeding season, at least according to K95%. For example, some studies reported larger home-range size in the breeding season [8], whereas others were reported in the non-breeding season [7,9]. These differences could be accounted for the different quality of information since our data are based on a larger sample size (n = 51) and higher accuracy (GPS/GSM technology versus radio-tracking or Argos-GPS information).

Furthermore, when "Season" and "Sex" were taken into account at the same time, we detected a smaller home-range size of females in the breeding season due to the lesser movement during the incubation and chicks' attendance. However, during the non-breeding season, no differences were found between males and females in their home-range. This could be explained by the fact that they spend most of their time together according to the GPS information and direct observation in the field (i.e., resting, cooperative hunting, flying, etc.).

4.2. Differences in Home-Range Size between Breeding Status and Seasons

The inflection point in territory size observed in June (Figure 2), where breeding individuals increase their home-range, coincides with the abandonment of the nest by juveniles. Once the juveniles are more independent, relaxation and less parental effort take place, and thus an increase in their home-range is observed. On the other hand, non-breeding individuals maintain constant territory throughout the year. Our results show that the home-range of breeding individuals increases, whereas those of non-breeding individuals decreases. This can be explained by the fact that breeding pairs are the ones that set the tone since they have more energy requirements and are presumably in better physical shape. In contrast, the non-breeding pairs are the ones that need to adjust to these requirements, occupying or ceding the territory, depending on the breeding pairs. Moreover, this inflection point is observed in December, where the opposite occurs, increasing the breeding area in non-breeding individuals and decreasing in the breeding ones.

4.3. Overlap between Neighboring Territories

We observed a low overlap of home-ranges between neighbors. This could be explained by the high level of intraspecific competition of Bonelli's eagle. The consistency of our results (4.18%) with previous studies is remarkable (4.1% in [7]; 4.33% in [8]). Bonelli's eagles are territorial raptors that exhibit agonistic behavior in the case of territorial invasion and they even extend their territory in the case of disappearance of a neighboring pair (authors' pers. obs.).

4.4. Implications for Conservation

The information provided in this study on the home-range size and ranging behavior of the Bonelli's eagle throughout the annual cycle provides a useful tool for the management and conservation of this species. Bonelli's eagle is one of the key species used to delimitate protected areas in Mediterranean ecosystems. Therefore, a detailed knowledge of the home-range size and its variation throughout the annual cycle could be used for a significantly effective design. Moreover, this information will make it possible to assess the impacts of various anthropic activities and to mitigate or even eliminate them. In contrast to the general pattern used decades ago, in which land planning was focused only on protecting nesting areas and the immediate nearby, our study shows that the continued presence of pairs throughout its territory year-round (see K95%, K75%, and K50%) requires a comprehensive protection or management of the entire home-range, including not only nesting areas, but also feeding and resting grounds.

Supplementary Materials: The following supporting information can be downloaded at: https://www.mdpi.com/article/10.3390/d14121082/s1. Figure S1: Monthly average of the daily home-range size at three different levels (K95%, K75%, and K50%). Figure S2: Boxplot of the monthly average of the daily home-range size (K95%) per territory. Figure S3: Differences in daily home-range size (K95%) among years. Figure S4: Differences in daily home-range size (K95%) among years and territories. Figure S5: Differences in daily home-range size (K95%) among years per individual. Figure S6: Differences in daily home-range size (K95%) between sexes. Figure S7: Differences in daily home-range size (K95%) between sexes and territories. Figure S8: Differences in daily home-range size (K95%) between seasons. Figure S9: Differences in daily home-range size (K95%) between seasons and territories. Figure S10: Differences in daily home-range size (K95%) between sexes, seasons, and territories. Figure S11: Differences in daily home-range size (K95%) between the breeding status. Figure S12: Differences in daily home-range size (K95%) between the breeding status and individuals. Table S1: Summary information of the 51 Bonelli's eagles tracked by GPS/GSM satellite telemetry in eastern Spain. Table S2: Summary statistics of daily home-range size (km^2) per individual according to three different spatial estimators (i.e., K95%, K75%, and K50%). Table S3: ANOVA results of interannual variation in monthly average daily home-range size according to the 95% kernel by individual. Significant variables are highlighted in bold.

Author Contributions: S.M., P.L.-L. and V.U. conceived the ideas, designed the methodology, collected the data, wrote the manuscript, and contributed critically to the drafts. S.M. and P.L.-L. analyzed the data. P.L.-L. and V.U. provided the final approval for publication. All authors have read and agreed to the published version of the manuscript.

Funding: This work was supported by Red Eléctrica de España, ACCIONA Eólica de Levante, LafargeHolcim and the Wildlife Service of the Valencian Community regional government (Conselleria d'Agricultura, Desenvolupament Rural, Emergència Climàtica i Transició Ecològica, Generalitat Valenciana, Spain).

Institutional Review Board Statement: Ethical review and approval were waived for this study because it does not include animals used for experimentation (see Ethical Statement).

Data Availability Statement: All data used in this study are publicly available upon request to data managers in the online data repository Movebank (www.movebank.org (accessed on 2 July 2022)). The projects are: "Bonelli's eagle University of Alicante Spain" (project ID = 58923588), "Bonelli's eagle Alicante Spain" (ID = 430140799), "Bonelli's eagle University of Valencia Spain" (ID = 193515984), and "Movement ecology of large raptors in Spain" (ID = 640908212).

Acknowledgments: We would like to thank F. García, J. Giménez, V. García, J. De la Puente, A. Bermejo, M. Montesinos, J.M. Lozano, M. Aguilar, M.A. Monsalve, A. Perona, A. López, I. Estellés, O. Egea, F. Cervera, J. Crespo, M. Vilalta, M. Surroca, T. De Chiclana, S. Ferreras, C. García, E. Mondragón, T. Camps, M. Marco, and V. Agustí for their help in fieldwork and eagles trapping. Special thanks to J. Jiménez of the regional government (Generalitat Valenciana's Wildlife Service) for his help with this project. This paper takes part in Sara Morollón's doctoral thesis at the University of Alicante.

Conflicts of Interest: The authors declare no conflict of interest.

References

1. BirdLife International. Species Factsheet: *Aquila fasciata*. 2022. Available online: http://www.birdlife.org (accessed on 2 July 2022).
2. Del Hoyo, J.; Elliot, A.; Sargatal, J. *Handbook of the Birds of the World*; Linx Edicions: Barcelona, Spain, 1994; Volume II, New world vultures to guineafowl.
3. López-López, P.; García-Ripollés, C.; Urios, V. Population Size, Breeding Performance and Territory Quality of Bonelli's Eagle Hieraaetus Fasciatus in Eastern Spain. *Bird Study* **2007**, *54*, 335–342. [CrossRef]
4. López-López, P.; Perona, A.; Egea-Casas, O.; Morant, J.; Urios, V. Tri-Axial Accelerometry Shows Differences in Energy Expenditure and Parental Effort throughout the Breeding Season in Long-Lived Raptors. *Curr. Zool.* **2022**, *68*, 57–67. [CrossRef] [PubMed]
5. Morollón, S.; Urios, V.; López-López, P. Fifteen Days Are Enough to Estimate Home-Range Size in Some Long-Lived Resident Eagles. *J. Ornithol.* **2022**, *163*, 849–854. [CrossRef]
6. Morollón, S.; López-López, P.; Urios, V. A New View of Territoriality in Large Eagles: The Territory Pre-Exists Regardless of Their Occupants. *Res. Sq.* **2022**. [CrossRef]
7. Bosch, R.; Real, J.; Tinto, A.; Zozaya, E.L.; Castell, C. Home-ranges and Patterns of Spatial Use in Territorial Bonelli's Eagles Aquila Fasciata. *IBIS* **2010**, *152*, 105–117. [CrossRef]

8. Pérez-García, J.M.; Margalida, A.; Afonso, I.; Ferreiro, E.; Gardiazábal, A.; Botella, F.; Sánchez-Zapata, J.A. Interannual Home Range Variation, Territoriality and Overlap in Breeding Bonelli's Eagles (*Aquila fasciata*) Tracked by GPS Satellite Telemetry. *J. Ornithol.* **2013**, *154*, 63–71. [CrossRef]
9. Martínez-Miranzo, B.; Banda, E.; Gardiazábal, A.; Ferreiro, E.; Aguirre, J.I. Differential Spatial Use and Spatial Fidelity by Breeders in Bonelli's Eagle (*Aquila fasciata*). *J. Ornithol.* **2016**, *157*, 971–979. [CrossRef]
10. Kenward, R.E. *A Manual for Wildlife Radio Tagging*; Academic Press: Cambridge, MA, USA, 2000.
11. García, V.; Iglesias-Lebrija, J.J.; Moreno-Opo, R. Null Effects of the Garcelon Harnessing Method and Transmitter Type on Soaring Raptors. *IBIS* **2021**, *163*, 899–912. [CrossRef]
12. Worton, B.J. Kernel Methods for Estimating the Utilization Distribution in Home-range Studies. *Ecology* **1989**, *70*, 164–168. [CrossRef]
13. Signer, J.; Balkenhol, N. Reproducible Home Ranges (Rhr): A New, User-friendly R Package for Analyses of Wildlife Telemetry Data. *Wildl. Soc. Bull.* **2015**, *39*, 358–363. [CrossRef]
14. R Core Team. *R: A Language and Environment for Statistical Computing*; R Core Team: Vienna, Austria, 2022.
15. Wells, R.; TinHan, T.C.; Dance, M.A.; Drymon, J.M.; Falterman, B.; Ajemian, M.J.; Stunz, G.W.; Mohan, J.A.; Hoffmayer, E.R.; Driggers, W.B., III. Movement, Behavior, and Habitat Use of a Marine Apex Predator, the Scalloped Hammerhead. *Front. Mar. Sci.* **2018**, *5*, 321. [CrossRef]
16. Miller, K.J.; Erxleben, D.R.; Rains, N.D.; Martin, J.C.; Mathewson, H.A.; Meik, J.M. Spatial Use and Survivorship of Translocated Wild-caught Texas Horned Lizards. *J. Wildl. Manag.* **2020**, *84*, 118–126. [CrossRef]
17. Osipova, L.; Okello, M.; Njumbi, S.; Ngene, S.; Western, D.; Hayward, M.; Balkenhol, N. Using Step-selection Functions to Model Landscape Connectivity for African Elephants: Accounting for Variability across Individuals and Seasons. *Anim. Conserv.* **2019**, *22*, 35–48. [CrossRef]
18. Montalvo, V.H.; Fuller, T.K.; Saénz-Bolaños, C.; Cruz-Díaz, J.C.; Hagnauer, I.; Herrera, H.; Carrillo, E. Influence of Sea Turtle Nesting on Hunting Behavior and Movements of Jaguars in the Dry Forest of Northwest Costa Rica. *Biotropica* **2020**, *52*, 1076–1083. [CrossRef]
19. Van Der Marel, A.; Waterman, J.M.; López-Darias, M. Social Organization in a North African Ground Squirrel. *J. Mammal.* **2020**, *101*, 670–683. [CrossRef]
20. Wysong, M.L.; Hradsky, B.A.; Iacona, G.D.; Valentine, L.E.; Morris, K.; Ritchie, E.G. Space Use and Habitat Selection of an Invasive Mesopredator and Sympatric, Native Apex Predator. *Mov. Ecol.* **2020**, *8*, 18. [CrossRef]
21. Rueda, C.; Jiménez, J.; Palacios, M.J.; Margalida, A. Exploratory and Territorial Behavior in a Reintroduced Population of Iberian Lynx. *Sci. Rep.* **2021**, *11*, 14148. [CrossRef]
22. Shakeri, Y.N.; White, K.S.; Waite, J.N. Staying Close to Home: Ecological Constraints on Space Use and Range Fidelity in a Mountain Ungulate. *Ecol. Evol.* **2021**, *11*, 11051–11064. [CrossRef]
23. Perona, A.M.; Urios, V.; López-López, P. Holidays? Not for All. Eagles Have Larger Home Ranges on Holidays as a Consequence of Human Disturbance. *Biol. Conserv.* **2019**, *231*, 59–66. [CrossRef]
24. Williams, S.M.; Lindell, C.A. The Influence of a Single Species on the Space Use of Mixed-Species Flocks in Amazonian Peru. *Mov. Ecol.* **2019**, *7*, 37. [CrossRef]
25. Liu, T.; Xu, Y.; Mo, B.; Shi, J.; Cheng, Y.; Zhang, W.; Lei, F. Home Range Size and Habitat Use of the Blue-Crowned Laughingthrush during the Breeding Season. *PeerJ* **2020**, *8*, e8785. [CrossRef]
26. Morant, J.; Abad-Gómez, J.M.; Álvarez, T.; Sánchez, Á.; Zuberogoitia, I.; López-López, P. Winter Movement Patterns of a Globally Endangered Avian Scavenger in South-Western Europe. *Sci. Rep.* **2020**, *10*, 17690. [CrossRef] [PubMed]
27. Abril-Colón, I.; Alonso, J.C.; Palacín, C.; Ucero, A.; Álvarez-Martínez, J.M. Factors Modulating Home Range and Resource Use: A Case Study with Canarian Houbara Bustards. *Mov. Ecol.* **2022**, *10*, 49. [CrossRef] [PubMed]
28. Wang, C.; Yu, X.; Xia, S.; Liu, Y.; Huang, J.; Zhao, W. Potential Habitats and Their Conservation Status for Swan Geese (*Anser cygnoides*) along the East Asian Flyway. *Remote Sens.* **2022**, *14*, 1899. [CrossRef]
29. Noonan, M.J.; Tucker, M.A.; Fleming, C.H.; Akre, T.S.; Alberts, S.C.; Ali, A.H.; Altmann, J.; Antunes, P.C.; Belant, J.L.; Beyer, D. A Comprehensive Analysis of Autocorrelation and Bias in Home Range Estimation. *Ecol. Monogr.* **2019**, *89*, e01344. [CrossRef]
30. Samuel, M.D.; Pierce, D.J.; Garton, E.O. Identifying Areas of Concentrated Use within the Home Range. *J. Anim. Ecol.* **1985**, *54*, 711–719. [CrossRef]
31. Kie, J.G.; Matthiopoulos, J.; Fieberg, J.; Powell, R.A.; Cagnacci, F.; Mitchell, M.S.; Gaillard, J.-M.; Moorcroft, P.R. The Home-Range Concept: Are Traditional Estimators Still Relevant with Modern Telemetry Technology? *Philos. Trans. R. Soc. B Biol. Sci.* **2010**, *365*, 2221–2231. [CrossRef]
32. Zuur, A.F.; Ieno, E.N.; Walker, N.J.; Saveliev, A.A.; Smith, G.M. *Mixed Effects Models and Extensions in Ecology with R*; Springer: Berlin/Heidelberg, Germany, 2009; Volume 574.
33. Harrison, X.A.; Donaldson, L.; Correa-Cano, M.E.; Evans, J.; Fisher, D.N.; Goodwin, C.E.; Robinson, B.S.; Hodgson, D.J.; Inger, R. A Brief Introduction to Mixed Effects Modelling and Multi-Model Inference in Ecology. *PeerJ* **2018**, *6*, e4794. [CrossRef]
34. Bates, D.; Mächler, M.; Bolker, B.; Walker, S. Fitting Linear Mixed-Effects Models Using Lme4. *J. Stat. Softw.* **2015**, *67*, 1–48. [CrossRef]
35. Hijmans, R.J.; van Etten, J. *Raster: Geographic Data Analysis and Modeling*, R Package Version 3.4-5. 2020. Available online: https://cran.r-project.org/web/packages/raster/raster.pdf (accessed on 2 July 2022).

36. Sanz, A.; Minguez, E.; Anadon, J.; Hernandez, V. Heterogeneous use of space in three breeding territories of Bonelli's eagle (*Hieraaetus fasciatus*). *Ardeola* **2005**, *52*, 347–350.
37. López-López, P. Individual-Based Tracking Systems in Ornithology: Welcome to the Era of Big Data. *Ardeola* **2016**, *63*, 5–34. [CrossRef]
38. Carrascal, L.M.; Seoane, J. Factors Affecting Large-Scale Distribution of the Bonelli's Eagle *Aquila fasciata* in Spain. *Ecol. Res.* **2009**, *24*, 565–573. [CrossRef]
39. Ferrer, M.; Morandini, V.; Newton, I. Floater Interference Reflects Territory Quality in the Spanish Imperial Eagle Aquila Adalberti: A Test of a Density-Dependent Mechanism. *IBIS* **2015**, *157*, 849–859. [CrossRef]
40. Martínez-Miranzo, B.; Banda, E.; Aguirre, J.I. Home Range Requirements in Bonelli's Eagle (*Aquila fasciata*): Prey Abundance or Trophic Stability? *Eur. J. Wildl. Res.* **2019**, *65*, 85. [CrossRef]

diversity

MDPI

Article

Foraging Patterns of Non-Territorial Eastern Imperial Eagle (*Aquila heliaca*): A Case of Successful Adaptation

Dimitar Demerdzhiev [1,2,*], Ivaylo Angelov [1,2,†] and Dobromir Dobrev [1]

1 Bulgarian Society for the Protection of Birds/BirdLife Bulgaria 5, Leonardo da Vinci Str., 4000 Plovdiv, Bulgaria
2 National Museum of Natural History, Bulgarian Academy of Sciences, 1000 Sofia, Bulgaria
* Correspondence: dimitar.demerdzhiev@bspb.org or dimitar.demerdzhiev@gmail.com
† Present address: Nature Conservation Centre Eastern Rhodopes, Madzharovo, 6300 Haskovo, Bulgaria.

Abstract: The Optimal Foraging Theory predicts that, to maximize fitness, animals adapt their foraging strategy that provides the most benefit for the lowest cost, maximizing the net energy gained. While the diet of many breeding raptor populations is well known, studies on the foraging patterns of non-territorial birds of prey (floaters) are scarce. In this study, we examined the foraging pattern of non-territorial Eastern Imperial Eagle, scrutinizing different aspects of its feeding ecology and behavior. We built a simple model of the optimal foraging strategy of floater eagles including the success of foraging as a currency as well as environmental factors such as seasons, type of prey, habitat, foraging techniques, and eagle age as a limitation affecting the foraging efficiency of birds. We found that floaters focused their diet exclusively on European Souslik, accounting for almost half (44.2%) of the eagle's prey. Diet differences between floaters and breeders were due to higher Souslik and carrion consumption and lower Hedgehog predation by floater eagles. The diet diversity of breeding eagles (H = 3.297) was much higher than that of floaters (H = 1.748). Our model suggested that the foraging mode, habitat type, and season best explained the feeding success of non-territorial eagles ($_{\Delta}$AIC = 0.00, w = 0.42). Of all explanatory factors, "Kleptoparasitism" ($\beta^2 = -4.35$), "Rodents" ($\beta^2 = -4.52$), "Pasture" ($\beta^2 = 2.96$), "Wheat" ($\beta^2 = 4.41$), "In the air" ($\beta^2 = 4.16$), and "Other habitats" ($\beta^2 = 4.17$) had a pronounced effect. The factors "Spring–summer season" ($\beta^2 = -0.67$) and "European Souslik" ($\beta^2 = -2.76$) had a marginal effect in our models. Generally, the mean success rate of attack modes used by non-territorial eagles was 0.54 ± 0.50. Floaters successfully obtained food through: kleptoparasitism (43.10%), carrion feeding (24.14%), and high soar with vertical stoop (14.66%). Several important issues for the conservation of non-territorial Eastern Imperial Eagles arose from our research. The strong relation of floaters with the European Souslik calls for specific conservation measures aimed at the conservation of this type of prey and the restoration and appropriate management of its grassland habitats. The importance of the scavenging behavior of juvenile birds requires increased control of the use of poison baits and subsequent prosecution by state institutions. Protecting the most important temporary areas, improving institutional control against the use of poison baits, and intensifying awareness-raising campaigns among pigeon-fanciers and hunters are also of crucial importance for effective species conservation.

Keywords: top predator; floaters; diet; raptors; feeding ecology; behavior; conservation; temporary areas

Citation: Demerdzhiev, D.; Angelov, I.; Dobrev, D. Foraging Patterns of Non-Territorial Eastern Imperial Eagle (*Aquila heliaca*): A Case of Successful Adaptation. *Diversity* **2022**, *14*, 1060. https://doi.org/10.3390/d14121060

Academic Editor: Huw Lloyd

Received: 8 November 2022
Accepted: 29 November 2022
Published: 2 December 2022

1. Introduction

The classic Optimal Foraging Theory (OFT) predicts that, to maximize fitness, animals adapt their foraging strategy that provides the most benefit (energy) for the lowest cost (time/effort), maximizing the net energy gained [1,2]. However, new concepts, such as balancing between foraging and safety, the assumption that tactical foraging decisions depend on state variables, such as fat reserves (state dependence theory), foraging games (game theory), and the consequences of foraging in a group, are incorporated in the decisions the animal must make [3]. The new aspect of the feeding behavior concerns the physiological, biochemical, and anatomical mechanisms that can constrain an animal and thus influence its foraging actions. Foraging behavior is crucial with regard to evolutionary biology not only because it is a major factor in the survival, growth, and reproduction of animals but also because of the resulting adaptations that persist in the course of evolution. This complex process is influenced by numerous factors from cognitive and physiological limitations to predation and social interactions [3,4].

While modeling foraging behavior, it should be kept in mind that organisms maximize a variable known as the currency, a unit including costs and benefits that are imposed on the animal. The constraints of the environment are key factors that can limit the forager's ability to maximize the currency. Then, the organism's best foraging strategy is defined as the decision that maximizes the currency under the constraints of the environment [5]. In most species, the availability and accessibility of food resources have been identified as the key factors that shape foraging behavior and dietary decisions [2,6]. Raptors can adapt their foraging as a response to main prey depletion [7] or to avoid competition [8].

The food spectrum and dietary relationship with breeding are well known for different diurnal and nocturnal birds of prey [9–14]. However, studies on the foraging patterns of non-territorial birds (floaters) of raptor species are scarce [15–19]. For many raptors, survival during the dispersal period has important consequences for the population trajectories in the future ([20,21]. During this period, the floaters of some large eagles tend to restrict their movements to a few favorable places, known as "temporary settlement areas" [20,22,23]. These sites are normally outside of the breeding territories and are characterized by the abundance of prey, where floaters spend periods of varying duration before joining breeding populations [15,22,24]. However, this differs from the juvenile high-mobility strategy demonstrated by tropical raptors [25].

The Eastern Imperial Eagle (*Aquila helica*) (hereafter EIE) is a large raptor species distributed from Eastern Europe to Central Asia [26,27]. This open-ground eagle forages in different habitats, where it exploits diverse prey such as Sousliks (*Spermophilus* sp.), Leporids (*Leporidae*), Hedgehogs (*Erinaceus* sp.), Corvids (*Corvidae*), Gulls (*Larus* sp.), White Stork (*Ciconia ciconia*), and various Reptiles (*Squamata, Testudinidae*) [10,11,28–34]. While the study of the diet of breeding EIEs is well documented throughout the distribution range of the species [10,30–32,34], the foraging of floaters is only known from sporadic observations [19], and there is no systematic survey presenting the various aspects of the foraging pattern of non-territorial birds. Although the populations of EIE in Europe are stable or even increasing in some parts of the distribution range [10,26], the species' global population is considered to be declining [35]. The species suffered from habitat loss and alteration, electrocution by power lines, poison baits, direct persecution, and prey decrease affecting both breeding territories and dispersal areas [31,36–38]. Therefore, preserving and securing appropriate settlement areas [21], along with the conservation of breeding grounds [36], is crucial for EIE population viability.

In this study, we examined the foraging patterns of non-territorial EIEs, exploring different aspects of their feeding ecology and behavior. By searching relationships between the success of feeding and the factors limiting/supporting it, we built a simple model of floater EIEs foraging including the success of foraging as a currency and environmental factors such as seasons, type of prey, habitat, foraging techniques, and eagle age as a constraint limiting the foraging efficiency of eagles. We predicted that: (1) eagles used different techniques to obtain food at different ages, and adults were more experienced and more successful in hunting, while less experienced juveniles and immature individuals could have a lower hunting success and, as a result, used alternative techniques (kleptoparasitism, scavenging) to optimize foraging; (2) habitat type, vegetation height, seasons, and type of prey influenced foraging strategy and success. We compared our data on floaters' diet with the information available for the breeding population and hypothesized that the diet of non-territorial eagles differed from the food of territorial birds and that this dietary diversification could be considered an example of successful adaption driven by evolution. Based on our findings, we discussed and recommended conservation efforts that should be taken to preserve and secure important settlement areas. Our results can be applied for other threatened large raptor species using the same foraging strategy and behavior during the dispersal period.

2. Materials and Methods

2.1. Study Design

We collected data from dispersal areas (Figure 1) of EIE in Bulgaria over a period of 25 years (1998–2022). A total of 186 cases of the foraging of non-territorial EIEs were documented, of which only 7 (3.76%) of the attempts were of unknown capture success due to the distance from the observer and/or local topography. Most of the observations (n = 147; 79%) were made in two important temporary settlement areas, Besaparski hills and Sliven field, which harbored dozens of floaters yearly [24,39]. These two areas were very similar and consisted of karst hills with almost no vegetation, as well as thermophilic grass communities dominated by Bluestem (*Dichanthium ischaemum*), Scented grass (*Chrysopogon gryllus*), and Needle grass (*Stipa capilata*), imparting the steppe character of the habitats [40]. In addition, we used the data (food remains; pellets, n = 22) of 20 identified preys of three EIEs equipped with satellite transmitters (PPT). Having located the roost sites of the birds, we visited the area and collected food remains and pellets.

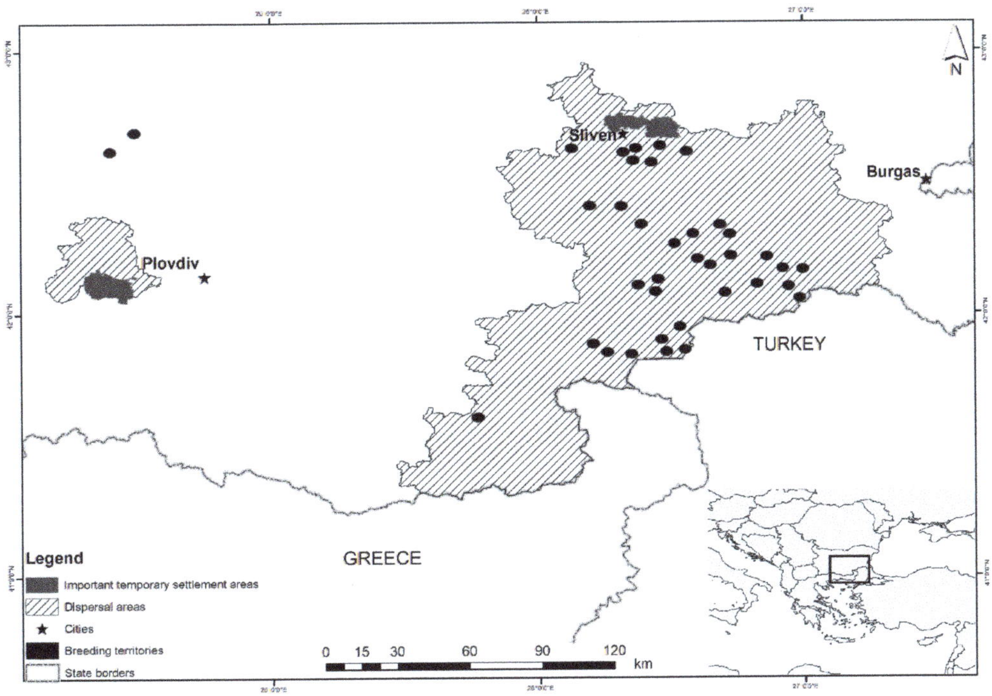

Figure 1. Dispersal areas (shaded) and breeding territories (black dots) used by the Eastern Imperial Eagle (*Aquila heliaca*). Two important temporary settlement areas (Besaparski hills and Sliven field) are given in grey.

Observations were made during the whole study period, which were particularly intense in both important settlement areas (Besaparski hills and Sliven field) during the period 2002–2011. During these years, the two sites were visited either two or three times per month, with an observation duration of two to eight hours. In the rest of the studied period, the two places were visited once every two or three months. The remaining observations were made in dispersal areas of secondary importance throughout the entire years. All observations in the study accounted for >1700 h of field work. For each observation, we noted the following information: day, hour, individual, eagle age, habitat characteristics, hunting or feeding technique used, success of the feeding/hunting technique, and prey [16]. Eagle plumage was identified according to [41]. For the purpose of the analyses, we grouped the different ages of eagles into three classes: juvenile (first calendar autumn—second calendar summer); immature (between second calendar autumn and sixth plumage); and adult (more than seven calendar autumn). An attack was defined as a direct attempt to capture/steal clearly identifiable prey [42], and a capture was an attack that resulted in the acquisition of prey by an eagle [43]. Thus, each attack was classified either as a success or a failure. Observations were made from points offering good visibility and consisted in active scans to detect the predator and the prey within a radius of 400–500 m around the observer [17]. All capture attempts with undetermined outcomes were excluded from the analysis [44]. However, all the observations of carcass feeding (n = 28) were considered successful because, in all cases, the individual accessed the carrion source [16].

The type of attack was defined by the position of the eagle at the beginning of the attack. Hunting behavior was determined according to [45,46] with some additions. We categorized five different hunting techniques to obtain food (Table 1). The other foraging techniques were separated in single categories. Six variables were evaluated for each attack (Table 1).

Table 1. Definition of the variables used in models to analyze the attack success and feeding behavior of Eastern Imperial Eagles.

Variable	Variable Type	Description
Age	Categorical	Three age-classes: adults, immature, and juveniles.
Attack/Foraging Mode	Categorical	Ten classes: (1) Powered contour flight; (2) High soar with vertical stoop and descent attack; (3) Glide attack with tail-chase; (4) High-perch; (5) Walk-grab; (6) Collect a crashed animal; (7) Kleptoparasitism; (8) Carrion feeding; (9) Cooperative feeding; (10) Unspecified
Prey type	Categorical	Prey species. The prey's single specimens are grouped into a common category. Nonidentified prey was given in a separate category.
Habitat type	Categorical	Habitat characteristics according to land use pattern: (1) Pasture; (2) Stubble; (3) Wheat; (4) Fallow; (5) Other, including single casses such as asfalt road, ekoton, quarry, shrubs, and fishpond;.(6) The airstrikes were divided into a separate category: "in the air"
Vegetation height	Continuous	The height of the vegetation in cm.
Seasons	Categorical	Two categories with an equal duration: (1) Spring–summer (from March to August) and (2) autum–winter (from September to February)

2.2. Statistical Analyses

We carried out a non-parametric Chi-square test with Monte Carlo randomization (9999 permutations) [47] to compare the diet of floaters in dispersal areas and breeding populations [15]. The data about the breeding EIE's diet were taken from the available literature [10]. However, the breeding population and the studied dispersal territory were distributed in the same area, with a maximum distance of 30 km between the dispersal places and breeding pairs (Figure 1). Therefore, we assume that our dietary data were geographically independent. Although different methods were used to collect the information (analysis of food remains and pellets for the breeding population vs. mostly visual observations of diet patterns of floaters), the two surveys were conducted simultaneously and covered the same annual/monthly periods. Therefore, the restrictions arising from the different methods are not significant and describe the general dietary patterns of the two sections of the population (breeders and floaters).

Food diversity calculated with the Shannon–Weaver index (H) [48] was also used in the diet analysis of the breeding population [10]. We applied Abundance-based Coverage Estimator (ACE) [49] for the bias correction (bootstrap 9999 simulations) of H. This corresponded to the Bias-corrected Maximum Likelihood Estimator (MLE) for Shannon's index given by [50].

To estimate the success rate of different foraging strategies used by eagles, we built a simple model using Generalized Linear Mixed Models (GLMM) with a binomial error distribution and logit link function. Foraging success was modeled as a binary variable (1 = success, 0 = failure). The constraint factors of foraging success included in the model as response variables were: season, age class of eagles, foraging mode, prey type, habitat type, and vegetation height (Table 1). Due to the impossibility of separate birds being individually identified and tracked over time, we did not include the "individual eagle ID" as a random factor in our model. Thus, the effect of the "individual" was not evaluated. We used the Akaike Information Criterion corrected for small sample sizes (AICc) for model selection and chose the models with the lowest AIC_c value from the set of our candidate models. All models with an AIC_c value < 2 from the model with the lowest AIC_c (AIC_{cmin}) were considered the best models ($_{\Delta}AIC_c = AIC_i - AIC_{cmin}$) [51]. The relative importance of each model was estimated through the weight of AICc (w), so all of the weights for all models added up to 1. Explanatory parameter estimates (β^2) with Lower (95%) and Upper CL (95%) and a probability value (p) of the explanatory factors were also evaluated.

All data were analyzed using Statistica for Windows, Release 12 [52], R v.2.15.2 [53], and Past Version 4.08 [54]). Results with $p \leq 0.05$ were considered significant. Values were provided as the means ± SE.

3. Results

3.1. Diet Diversity and Comparison of Floaters vs. Breeders

We found that floaters based their diet exclusively on European Souslik (*Spermophilus citellus*), accounting for almost half (44.2%) of the eagle's prey (Table 2). The other important food sources for non-territorial eagles were carrion (21.1%), followed by Small Rodents (Rodentia, excl. Souslik) (7.25%), Feral Pigeon (*Columba livia f. domestica*) (6.52%), and Brown Hare (*Lepus europaeus*) (3.62%). Other prey such as Rook (*Corvus frugilegus*), Golden Jackal (*Canis aureus*), Snakes (Serpentes), Common Kestrel (*Falco tinnunculus*), and Northern White-breasted Hedgehog (*Erinaceus roumanicus*) appeared occasionally (Table 2). Non-detected prey accounted for 10.14% of the consumed animals.

Table 2. Diet of non-territorial Eastern Imperial Eagle in dispersal areas.

Prey	Observation (ind.)	Food Remains (ind.)	Total (ind.)	Total (%)
Spermophilus citellus	50	11	61	44.20
Rodentia (excl. Souslik)	10		10	7.25
Columba livia f. domestica	1	8	9	6.52
Lepus europaeus	5		5	3.62
Serpentes	1	1	2	1.45
Falco tinnunculus	1		1	0.72
Carrion	29		29	21.01
Unidentified	14		14	10.14
Corvus frugilegus	4		4	2.90
Erinaceus roumanicus	1		1	0.72
Canis aureus	2		2	1.45
Total	**118**	**20**	**138**	**100**

Prey frequency in the diet significantly differed between floaters (in dispersal areas) and breeders (territorial pairs) due to higher Souslik ($\chi^2 = 9.94$, df = 1, Monte Carlo: $p = 0.002$) and carrion consumption ($\chi^2 = 16.22$, df = 1, Monte Carlo: $p < 0.001$) and lower Hedgehog ($\chi^2 = 23.62$, df = 1, Monte Carlo: $p < 0.001$) predation by floater eagles (Figure 2). The diet diversity of breeders (H = 3.297) was much higher than that of floaters (H = 1.748).

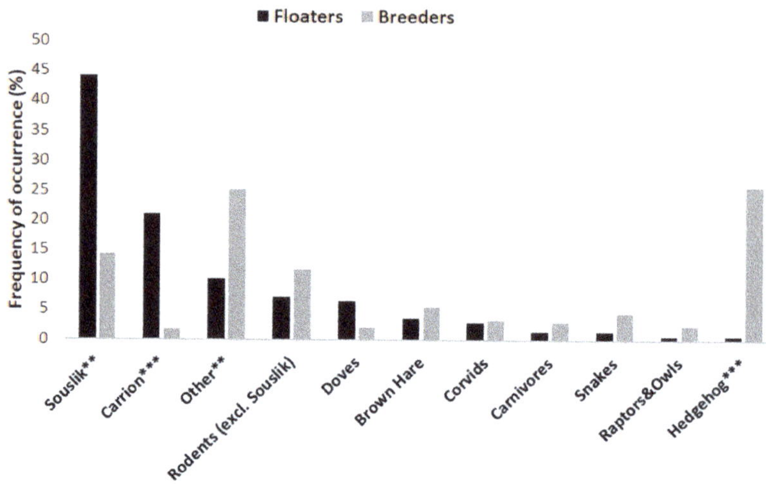

Figure 2. Comparison of prey occurrence in the diet of floaters and breeders of Eastern Imperial Eagle in Bulgaria. Data for breeders (territorial pairs) were taken from [10]. Significant values are given in *.

3.2. Foraging Pattern of Non-Territorial EIEs

By modeling the foraging pattern of non-territorial eagles, we found that "Seasons" + "Foraging Mode" + "Habitat type" primarily explained successful feeding ($_\Delta$AIC = 0.00, w = 0.42). The second-ranked model included only "Foraging Mode" and "Habitat type" ($_\Delta$AIC = 1.39, w = 0.21), while the third one included "Prey type" + "Habitat type" ($_\Delta$AIC = 1.50, w = 0.20) (Table 3). Of the explanatory factors, "Kleptoparasitism" ($\beta^2 = -4.35$, Wald. Stat. = 6.03, $p = 0.01$) and habitats had a pronounced effect (Table 3). Factors such as "Spring–summer season" ($\beta^2 = -0.67$, Wald. Stat. = 3.82, $p = 0.051$) and "European Souslik" ($\beta^2 = -2.76$, Wald. Stat. = 3.87, $p = 0.049$) had only a marginal impact in our models.

Table 3. List of GLMMs used for the analysis of the foraging pattern of non-territorial Eastern Imperial Eagle. All models with $_\Delta$AIC < 2 were considered the best models. The degree of freedom (df), model weight value (w) and probability value of each model (p) were also given. Parameter estimates (β^2) ± SE, Lower (95%) and Upper CL (95%) of significant explanatory factors, their importance value (Wald Stat.), and a probability value (p) were taken from the average model.

N	Model Structure	AIC	$_\Delta$AIC	df	w	p
1	Seasons + Foraging Mode + Habitat type	175.77	0.00	13	0.42	<0.001
2	Foraging Mode + Habitat type	177.16	1.39	12	0.21	<0.001
3	Prey type + Habitat type	177.26	1.50	20	0.20	<0.001
4	Seasons + Foraging Mode + Habitat type + Vegetation (cm)	177.43	1.66	14	0.17	<0.001
N	**Explanatory factors**	**β^2**	**St.err.**	**Lower CL/Upper CL**	**Wald Stat.**	**p**
1	Spring–summer season	−0.67	0.34	−1.34/0.00	3.82	0.051
2	Kleptoparasitism	−4.35	1.77	−7.81/−0.88	6.03	0.01
3	European Souslik	−2.76	1.40	−5.51/−0.01	3.87	0.049
4	Rodents (excl. Souslik)	−4.52	1.78	−8.02/−1.03	6.43	0.01
5	Pasture	2.96	0.54	1.90/4.02	30.07	<0.001
6	In the air	4.16	0.80	2.60/5.72	27.29	<0.001
7	Other habitats	4.17	1.05	2.12/6.22	15.87	<0.001
8	Wheat	4.31	0.67	3.00/5.62	41.46	<0.001

3.3. Foraging Mode and Success

Non-territorial EIEs successfully obtained food through: kleptoparasitism in 50 (43.10%) of the cases, high soar with vertical stoop in 17 (14.66%) of the successful attacks, walk-grab in 5 (4.31%) of the cases, cooperative feeding and high-perch in 2 cases each (1.72%), crashed animal pickup in 1 case (0.86%), and carrion feeding in 28 (24.14%) of the cases. In 11 (9.48%) of the cases, the foraging mode was unspecified.

The most successful attack techniques were walk-grab (n = 8 cases; 62.5% success) and kleptoparasitism (n = 81 cases; 61.73%), followed by high-perch (n = 4 cases, 50% success) and high soar with vertical stoop (n = 36 cases; 47.22% success). However, the hunting techniques of a proven unsuccessful rate were: glide attack with tail-chase (n = 7 cases) and powered contour flight in one case.

3.4. Effect of Habitat Type, Prey Type, and Season

As our global model demonstrated, pasture (n = 80), air habitat (n = 51), and wheat (n = 23) were the most common habitat types used by eagles. Fallow (90%), pasture (78.75%), and stubble (75%) were the habitats most successfully used for foraging by birds, while airily environments were mostly unsuccessful (60.78%). The attack success in wheat and other habitats was almost equal to unsuccessful attempts (Figure 3).

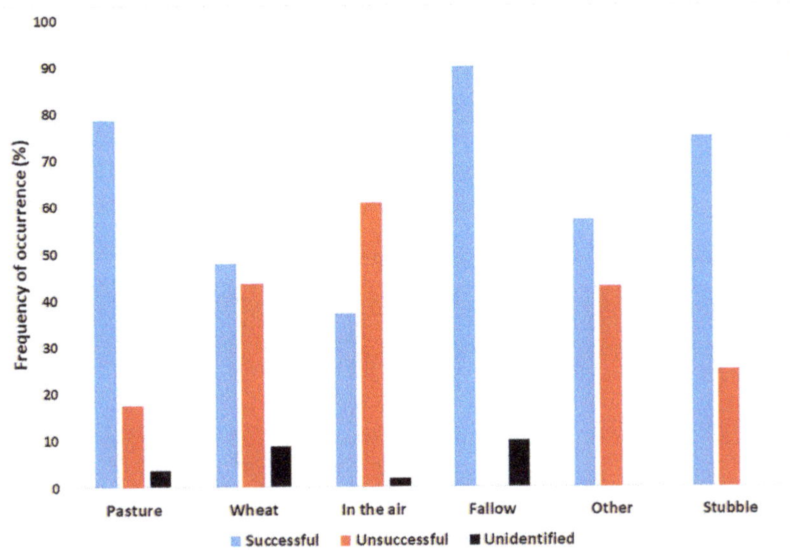

Figure 3. Results of the foraging of non-territorial Eastern Imperial Eagles in different habitat types.

In the spring–summer season, non-territorial EIEs exclusively exploited European Souslik (73.61%), while in the autumn–winter period, they consumed primarily carrion (42.31%) (Figure 4). However, in winter, Feral Pigeon (χ^2 = 11.66, df = 1, Monte Carlo: $p < 0.001$), Brown Hare (χ^2 = 4.37, df = 1, Monte Carlo: $p = 0.008$), and Rodents (χ^2 = 4.77, df = 1, Monte Carlo: $p = 0.01$) significantly increased their frequency in the eagle's diet (Figure 4). Regarding the active hunting techniques, the eagles used glide attack with tail-chase more frequently in winter (χ^2 = 5.54, df = 1, Monte Carlo: $p = 0.004$), while high soar with vertical stoop was used mostly in summer, although this difference was marginal (χ^2 = 3.05, df = 1, Monte Carlo: $p = 0.06$). However, 59.30% of the active hunting modes were successful in summer, while only 45.10% of the attacks benefited in the winter season.

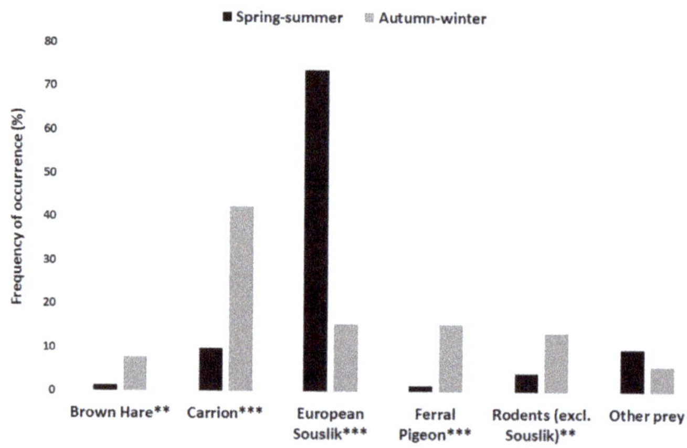

Figure 4. Frequency of occurrence (%) of different prey types used by non-territorial Eastern Imperial Eagles in different seasons. Significant values are given in *.

3.5. *Effect of Eagle Age*

As a whole, the mean success rate of attack modes used by non-territorial eagles was 0.54 ± 0.50 (n = 138). Corresponding with our "Global model", eagle age did not influence the success of attacking modes (Kruskal–Wallis test: $H_2 = 1.54$, $p = 0.46$). The mean success rate of adults was 0.48 ± 0.11 (n = 23). Immature eagles successfully attacked/stole prey with mean rate of 0.57 ± 0.05 (n = 105), while for juveniles, this rate was 0.40 ± 0.16 (n = 10). Surprisingly, juvenile eagles used kleptoparasitism in only 25% of the documented foraging (n = 20), and this significantly differed ($\chi^2 = 8.33$, df = 1, Monte Carlo: $p = 0.004$) from the registered cases for immature eagles (50%, n = 130) and adults (48%, n = 25) (Figure 5). Expectedly, juveniles relied mainly on carrion feeding (50% of foraging behavior), and this differed significantly ($\chi^2 = 21.61$, df = 1, Monte Carlo: $p < 0.0001$) from the scavenging practices of immature eagles (13.08%) and adults (4%) ($\chi^2 = 39.19$, df = 1, Monte Carlo: $p < 0.0001$). High soar with vertical stoop was the most often used active hunting technique for all age classes (Figure 5).

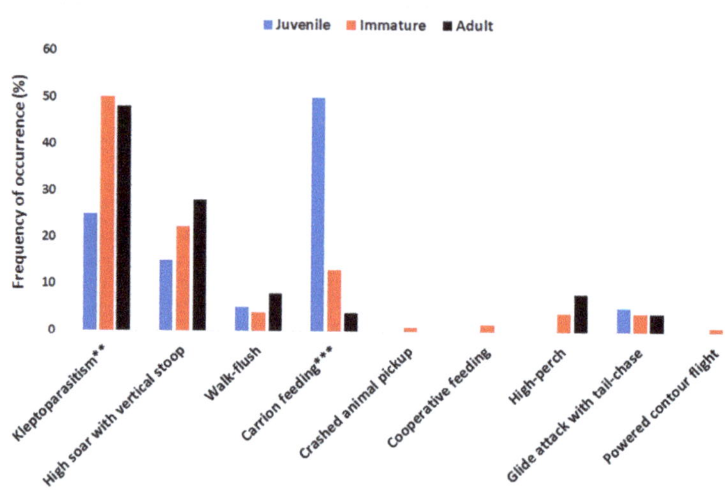

Figure 5. Frequency of occurrence (%) of different foraging modes used by different age classes of non-territorial Eastern Imperial Eagles. Significant values are given in *.

4. Discussion

4.1. Dietary Diversification as a Case of Successful Adaptation

Little is known about the foraging strategy and diet differences between floaters and territorial birds of large raptor species and how they choose prey and perform feeding behavior (however, see [15]). It is well known that the availability and accessibility of food resources are key factors that shape the foraging behavior and dietary choice [2,6]. Top predators can adapt their diet mainly in response to habitat alteration and the depletion of main food resources [7,55,56], but also to avoid competition [8]. Our hypothesis that the diet of non-territorial EIE differed significantly from the food choice of breeding birds was fully confirmed. Floater birds displayed dietary preferences for European Souslik, which largely determined their dispersal pattern and foraging strategy. In their dispersal grounds, they found temporary areas, where Souslik abundance existed, and where they formed concentrations by dozens. Searching for food, non-territorial eagles moved from site to site, sometimes covering great distances, where they opportunistically used any readily available food source such as carrion. Specialization in Souslik as a keystone prey led to low trophic diversity in non-territorial eagles. Sousliks had become ideal prey for inexperienced floater eagles, since they can be locally abundant, offering high energy value and low hunting costs. Consequently, floater EIEs depended heavily on Sousliks, and this would explain their ecology and behavior during the dispersal phase. In contrast, breeding EIEs depended on the type of prey in their territories, and when there were not enough Sousliks, they had to make a trade-off decision about whether to occupy that territory and adapt to other less profitable prey species, such as hedgehogs, or harder to catch prey, such as storks, or seek another territory, behaving like floaters, but this could hardly be proven. With an insufficient abundance of the main prey in the breeding territory, eagles exploited various type of caught prey and thus had a more diverse diet [10]. Similar diet diversification between floaters and breeders was found for another top predator: the Bonelli's Eagle (*Aquila fasciata*) [15].

We speculate that this diet difference between floaters and breeders represents an adaptive mechanism to avoid intraspecific competition in populations of large raptor species. Floaters, which are mostly younger and less experienced individuals, are attracted to places that have an abundance of easier-to-obtain and more profitable prey, such as Sousliks, usually away from territory defenders and non-tolerant breeders. Here, floaters more easily survive, increasing their skills, which would help them capture more difficult and diverse prey when they occupy a territory in the future. This phenomenon in the life history of eagles is a process probably driven by evolution, but deeper insight is needed.

4.2. Factors Influencing Foraging Behavior and Success

Our model suggests that foraging mode, habitat type, and season best explain feeding success, followed by prey type. Kleptoparasitism was the most effective hunting technique used by non-territorial EIEs, as almost half of the successful hunts consisted of stealing the prey from another predator. Contrary to our expectations, juvenile eagles quite rarely used kleptoparasitism as a foraging mode, unlike immature and adult birds, and this differed from the findings of another study on the Spanish Imperial Eagle (*Aquila adalberti*) [16]. Apparently, in a hierarchical relationship, juveniles are less experienced than older and more powerful eagles; hence, they try to steal prey from them less often, resorting exclusively to feeding on carrion. Conversely, more experienced and more suspicious immature and adult birds resort to the riskier and more atypical scavenging practices less often and use kleptoparasitism or active hunting. We consider that the differences in the findings of the Spanish Imperial Eagle study are due to the different methodological approach. Margalida et al. [16] compared the foraging mode of eagles of different age classes, including breeders and floaters, while our study was focused only on non-territorial birds. However, kleptoparasitic behavior is widespread among birds [57,58] and common among raptors [59]. The use of kleptoparasitism by floaters as a main foraging technique to obtain food probably differs from the foraging behavior of territorial eagles. Breeders must defend a territory, engage in mating behavior, and raise offspring, and these factors are more important in terms of maximizing fitness [2,60]. They have to minimize the time spent on foraging and probably do not have enough time, unlike floaters, which can sit and watch or soar for hours and wait for a convenient opportunity to steal prey from another predator. Then, kleptoparasitism seems to be a low-cost, highly profitable foraging mode in floater eagles. Anyway, this issue needs further clarification.

High soar with vertical stoop was the active hunting technique most frequently used by the eagles, yielding success in almost half of the cases. Birds used high soar primarily during the summer period, when ascending thermals particularly favored this type of prey searching. However, this hunting technique is widespread among many diurnal birds of prey [27]. Through walk and grab, eagles successfully obtained medium-sized or small prey such as Sousliks or voles. This hunting technique was frequently used by another eagle such as the Lesser Spotted Eagle (*Clanga pomarina*) in areas with a high prey density [61]. Glide attack with tail-chase was used by eagles exclusively during the winter period, when they tried to catch agile and maneuverable prey, such as pigeons, in the air. The predator's poor position when starting the attack, as well as the lack of surprise, were probably the reasons why all these attempts ended in failure. We found that the other hunting strike techniques, such as high-perch and powered contour flight, were rarely used by non-territorial EIEs. High-perch was a common hunting mode used by different large raptors such as the White-tailed Eagle (*Haliaeetus albicilla*), Spanish Imperial Eagle, Lesser Spotted Eagle, and Bonelli's Eagle [16,17,62,63]. However, powered contour flight with short glide attack and walk-grab were successfully used by another species closely related to EIE: the Steppe Eagle (*Aquila nipalensis*) [27].

In line with our expectations, the habitat type strongly affected the foraging success of eagles. This was in agreement with several studies that had shown that the landscape characteristics and prey types were also factors influencing hunting success [16,17,63]. While pasture, fallow, and stubble were the most successfully used habitats, the air environment was mostly unsuccessful. Being larger and less mobile, the EIE hunts mostly terrestrial prey dominated by various medium-sized and small mammals and reptiles. Birds on the ground are also easier to catch, especially when taken by surprise. Back in the air, eagles are not as agile and fast as hawks or falcons, and hunting there more often ends in failure. An important fact to consider here is that non-territorial eagles are mostly juvenile and immature birds, and hunting airborne prey requires more skills and experience. The small differences in the shape and proportions of wings and tails between immature and adult eagles also affect the hunting ability [27]. Feeding on Storks or other more difficult aerial prey, such as gulls and pigeons, has been well documented for territorial EIEs [10,30]. Apparently, after gaining experience, eagles adapt to hunting more difficult-to-capture aerial prey. The individual abilities are not unimportant either. However, whether territorial eagles specialize in hunting birds by capturing them in the air or primarily on the ground is a question that needs further investigation.

Our expectation that vegetation height influenced the strike success of eagles did not find support in this study. We assume that this was due to the type of habitats used. The pastures, where the birds were primarily observed hunting, were well managed with a low grass height. The other used habitats, such as stubble or fallow, were also characterized by low vegetation or by a total absence of greenery. Interestingly, in tall and dense wheat, where it was more difficult to hunt, we saw almost equal success and failure in hunting. The lack of sufficient instances of hunting in taller vegetation was probably due to the avoidance of this type of landscape by the eagles. More light is needed, however, to clarify this issue.

Corresponding to other studies [10,63], our research demonstrated a clear relationship between seasons and the type of taken prey. In the spring–summer season, European Souslik predominated in the diet, while in the autumn–winter period, eagles tried to compensate for the lack of Sousliks by feeding on carrion, Ferral Pegion, Brown Hare, and Small Rodents. However, in winter, due to deteriorating weather conditions, the hunting success was also lower. Similar results were found for another large top-predators such as the White-tailed Eagle [63].

The mean success rate of the attack modes found in our studies was similar to that recorded for White-tailed Eagle (50.5%) and greater than the one observed for Bonelli's Eagle (28.2%), Lesser Spotted Eagle (24%), and Golden Eagle (*Aquila chrysaetos*) (20%) [44,62,63]. Surprisingly, eagle age did not influence the success of attacking modes, and, at first sight, immature eagles had an even higher mean success rate than adults. This contradiction was due to the smaller sample size of adults foraging and the fact that immature eagles used mostly kleptoparasitism to obtain food. If we ignore kleptoparasitism as a phenomenon and consider only cases of active hunting, then the success of adult eagles would be 54.55%, and that of immature birds and juveniles would be 40%. This better corresponds to the other findings for different birds of prey [16,63–65]. Age-related improvements in foraging skills and experience benefit adults in using various hunting techniques, such as high soar with vertical stoop, walk-grab, and high-perch, more often than for the other age groups. An important fact to consider here is also the extent to which juvenile plumage morphology limits their ability to hunt the way that adults do [66]. The small differences in the length and stiffness of the remiges and rectrices between juveniles/immatures and adults put young birds at a disadvantage in terms of flying expertise [27]. Bulges of secondary feathers, shorter tails, and possibly softer remiges could hinder the ability of the juveniles to use high soar with vertical stoop and powered contour flight for hunting sousliks the way that adults do. However, neurological maturation and less developed pectoral muscles could also affect age differences in foraging behavior. Thus, further research is needed to better understand the age-related differences in the foraging behavior of large eagles.

4.3. Conservation Suggestion and Perspectives

Although some studies have investigated the diet of the breeding populations of EIEs in the entire distribution area [10,11,29–34], detailed reports during the dispersal phase are scarce, and this ecological trait during that particular life stage has not been considered in the management strategies for this threatened raptor.

Several important implications for EIE conservation emerge from our study. First of all, the strong dependence of non-territorial eagles on the European Souslik calls for specific conservation measures for this increasingly declining species [67]. There has been a trend for plowing and converting vast areas of grassland and semi-natural grass habitats into cropland over the last decade [68]. This process severely affects some of the most important places, both for the floater section of the EIE population and the breeding grounds. The floaters' tendency to form large concentrations in certain temporary settlement areas, where they spend long periods, requires the special protection of these places. Appropriate settlement areas may increase floaters' survival and guarantee population viability [21,69]. The recovery of Souslik populations through restocking [70], as well as the proper management of grassland habitats (through grazing by small animals such as sheep and goats) in these areas that are so important to non-territorial eagles, are of paramount importance for the conservation of the species. The significance of floaters to raptor population trajectories is well documented [21]. The most important places should be put under protection. The restoration of already damaged and plowed grassland habitats is also recommended.

Secondly, the importance of carrion for the survival of juvenile birds poses a severe risk of poisoning, due to illegal baits occasionally being used to control predators. Poisoning has been identified as the most important mortality factor affecting the breeding population of EIE in Bulgaria [37].

Thirdly, eagles feeding on pigeons can also raise conflicts with pigeon fanciers, which, in turn, could result in persecution incidents. Unfortunately, there have been such examples concerning non-territorial EIE [71].

Therefore, protecting the most important temporary settlement areas for non-territorial EIEs, restoring and subsequently properly managing damaged grassland habitats, strengthening the European Souslik through restocking programs, improving the monitoring of important large and/or threatened souslik colonies, preventing poisoning incidents through increased control of the use of poison baits, as well as intensifying awareness campaigning among key stakeholders, such as pigeon-fanciers and hunters, are paramount to the conservation of non-territorial EIEs.

5. Conclusions

Our study demonstrates that floaters of a large top predator, such as the EIE, generally adapt their diet via Souslik dominance, thus avoiding intraspecies competition by breeders, and this prey dependence influences their dispersal pattern. Eagles can modify their foraging strategy to cope with variations in weather conditions and food availability. Our predictions that the different eagle ages involved different techniques to obtain food and that the habitat type and prey type influenced the foraging success found clear support. Kleptoparasitism was the most successful mode for non-territorials to obtain food, while glide attack with tail-chase was fully unsuccessful.

Several important issues for the conservation of non-territorial EIEs arose from our research. The strong relation of floater eagles to European Souslik calls for specific conservation measures aimed at the conservation of this prey and the restoration and appropriate management of its grassland habitats. The importance of the scavenging behavior of juvenile birds requires increased control over the use of poison baits and subsequent prosecution by state institutions. Increased awareness campaigning among pigeon-fanciers and hunters is also of crucial importance for effective EIEs conservation.

Author Contributions: Conceptualization, D.D. (Dimitar Demerdzhiev) and I.A.; methodology, D.D. (Dimitar Demerdzhiev) and I.A.; software, D.D. (Dimitar Demerdzhiev) and D.D. (Dobromir Dobrev); validation, D.D. (Dimitar Demerdzhiev), I.A. and D.D. (Dobromir Dobrev); formal analysis, D.D. (Dimitar Demerdzhiev); investigation, D.D. (Dimitar Demerdzhiev), I.A. and D.D. (Dobromir Dobrev); resources, D.D. (Dimitar Demerdzhiev) and I.A.; data curation, D.D. (Dimitar Demerdzhiev), I.A. and D.D. (Dobromir Dobrev); writing—original draft preparation, D.D. (Dimitar Demerdzhiev); writing—review and editing, I.A. and D.D. (Dobromir Dobrev); visualization, D.D. (Dimitar Demerdzhiev) and D.D. (Dobromir Dobrev); supervision, D.D. (Dimitar Demerdzhiev) and D.D. (Dobromir Dobrev); project administration, D.D. (Dimitar Demerdzhiev); funding acquisition, D.D. (Dimitar Demerdzhiev) and I.A. All authors have read and agreed to the published version of the manuscript.

Funding: This work was partially funded by the projects: "Conservation of globally important biodiversity in high nature value semi-natural grasslands through support for traditional local economy" (GEFSEC Project No 43595) and "Conservation of Imperial Eagle and Saker Falcon in key Natura 2000 sites in Bulgaria" (LIFE07 NAT/BG/000068).

Institutional Review Board Statement: Not applicable.

Data Availability Statement: Not applicable.

Acknowledgments: We would like to express our most sincere gratitude to Vladimir Trifonov, Atanas Demerdzhiev, Stoycho Stoychev, Girgina Daskalova, Nikolay Terziev, Georgi Pogeorgiev, Dimitar Plachiyski, Vladimir Dobrev, Kiril Metodiev, Nedko Nedialkov, Krasimira Demerdzhieva, Volen Arkumarev, Valentin Gochev, Georgi Georgiev, Vera Dyulgerska, Hristo Hristov, Hristo Ivanov, Svetoslav Spasov, Stefan Avramov, Marin Kurtev, and Milan Bakalov, who took part in the field work. Without their assistance, this survey would not be possible. We are grateful to tree anonymous reviewers who improved the draft of the manuscript.

Conflicts of Interest: The authors declare no conflict of interest.

References

1. Pyke, G.; Pulliam, H.R.; Charnov, E.L. Optimal Foraging: A Selective Review of Theory and Tests. *Q. Rev. Biol.* **1977**, *52*, 137–154. [CrossRef]
2. Stephens, D.W.; Krebs, J.R. *Foraging Theory*; Princeton University Press: Princeton, NJ, USA, 1986.
3. Stephens, D.W.; Brown, J.S.; Ydenberg, R.C. (Eds.) *Foraging: Behavior and Ecology*; University of Chicago Press: Chicago, IL, USA, 2008.

4. Kramer, D.L. 'Foraging Behavior'. In *Evolutionary Ecology: Concepts and Case Studies*; Fox, C.W., Roff, D.A., Fairbairn, D.J., Eds.; Oxford Academic: Chenango, NY, USA, 2001. [CrossRef]
5. Sinervo, B. Optimal Foraging Theory: Constraints and Cognitive Processes. In *Behavioral Ecology*; Chapter 6; University of Santa Cruz: Santa Cruz, CA, USA, 1997; pp. 105–130.
6. Naef-Daenzer, L.; Naef-Daenzer, B.; Nager, R.G. Prey selection and foraging performance of breeding Great Tits *Parus major* in relation to food availability. *J. Avian Biol.* **2000**, *31*, 206–214. [CrossRef]
7. Moleón, M.; Sánchez-Zapata, J.A.; Real, J.; García-Charton, J.A.; Gil-Sánchez, J.M.; Palma, L.; Bautista, J.; Bayle, P. Large-scale spatio-temporal shifts in the diet of a predator mediated by an emerging infectious disease of its main prey. *J. Biogeogr.* **2009**, *36*, 1502–1515. [CrossRef]
8. Gryz, J.; Krauze-Gryz, D. Food Niche Overlap of Avian Predators (Falconiformes, Strigiformes) in a Field and Forest Mosaic in Central Poland. *Animals* **2021**, *11*, 479. [CrossRef] [PubMed]
9. Clouet, M.; Gerard, J.-F.; Goar, J.-L.; Goulard, M.; González, L.; Rebours, I.; Faure, C. Diet and Breeding Performance of the Golden Eagle *Aquila Chrysaetos* at the Eastern and Western Extremities of the Pyrenees: An Example of Intra-Population Variability. *Ardeola* **2017**, *64*, 347–361. [CrossRef]
10. Demerdzhiev, D.; Boev, Z.; Dobrev, D.; Terziev, N.; Nedyalkov, N.; Stoychev, S.; Petrov, T. Diet of Eastern Imperial Eagle (*Aquila heliaca*) in Bulgaria: Composition, distribution and variation. *Biodivers. Data J.* **2022**, *10*, e77746. [CrossRef]
11. Horváth, M.; Szitta, T.; Firmánszky, G.; Solti, B.; Kovács, A.; Moskát, C. Spatial variation in prey composition and its possible effect on reproductive success in an expanding Eastern Imperial Eagle (*Aquila heliaca*) population. *Acta Zool. Acad. Sci. Hung.* **2010**, *56*, 187–200.
12. Katzner, T.; Bragin, E.; Knick, S.; Smith, A. Relationship between demographics and diet specificity of Eastern Imperial Eagle *Aquila heliaca* in Kazakhstan. *IBIS* **2005**, *147*, 576–586. [CrossRef]
13. Lourenço, R.; Delgado, M.D.M.; Campioni, L.; Korpimäki, E.; Penteriani, V. Evaluating the influence of diet-related variables on breeding performance and home range behaviour of a top predator. *Popul. Ecol.* **2015**, *57*, 625–636. [CrossRef]
14. Murgatroyd, M.; Avery, G.; Underhill, L.; Amar, A. Data from: Adaptability of a specialist predator: The effects of land use on diet diversification and breeding performance of Verreaux's eagles. *J. Avian Biol.* **2016**, *47*, 001–012. [CrossRef]
15. Caro, J.; Ontiveros, D.; Pleguezuelos, J.M. The feeding ecology of Bonelli's eagle (*Aquila fasciata*) floaters in southern Spain: Implications for conservation. *Eur. J. Wildl. Res.* **2011**, *57*, 729–736. [CrossRef]
16. Margalida, A.; Colomer, M.À.; Sánchez, R.; Sánchez, F.J.; Oria, J.; González, L.M. Behavioral evidence of hunting and foraging techniques by a top predator suggests the importance of scavenging for preadults. *Ecol. Evol.* **2017**, *7*, 4192–4199. [CrossRef] [PubMed]
17. Martínez, J.E.; Zuberogoitia, I.; Gómez, G.; Escarabajal, J.M.; Cerezo, E.; Jiménez-Franco, M.V.; Calvo, J.F. Attack success in Bonelli's Eagle Aquila fasciata. *Ornis Fenn.* **2014**, *91*, 67–78.
18. Moleón, M.; Bautista, J.; Sánchez-Zapata, J.A.; Gil-Sánchez, J.M. Diet of non-breeding Bonelli's EaglesHieraaetus fasciatusat settlement areas of southern Spain. *Bird Study* **2009**, *56*, 142–146. [CrossRef]
19. Zhelev, P.; Gradev, G.; Ivanov, I.; Georgiev, D. On the food spectrum of juvenile Eastern Imperial Eagles (*Aquila heliaca* Savigny, 1809) in Southern Bulgaria. *Ecol. Balk.* **2009**, *1*, 51–58. (In Bulgarian)
20. Morrison, J.L.; Wood, P.B. Broadening Our Approaches to Studying Dispersal in Raptors. *J. Raptor Res.* **2009**, *43*, 81–89. [CrossRef]
21. Soutullo, A.; López-López, P.; Urios, V. Incorporating spatial structure and stochasticity in endangered Bonelli's eagle's population models: Implications for conservation and management. *Biol. Conserv.* **2008**, *141*, 1013–1020. [CrossRef]
22. Balbontín, J. Identifying suitable habitat for dispersal in Bonelli's eagle: An important issue in halting its decline in Europe. *Biol. Conserv.* **2005**, *126*, 74–83. [CrossRef]
23. Ferrer, M. *The Spanish Imperial Eagle*; Lynx Edicions: Barcelona, Spain, 2001; 224p.
24. Demerdzhiev, D. Eastern Imperial Eagle (*Aquila heliaca heliaca* Savigny, 1809) (Accipitridae—Aves) in Bulgaria—Distribution, Biology, Ecology, Numbers, and Conservation Measures. Synopsis of the Ph.D. Thesis, BAS-NMNH, Sofia, Bulgaria, 2011. (In Bulgarian, with English Summary)
25. Zuluaga, S.; Vargas, F.H.; Aráoz, R.; Grande, J.M. Main aerial top predator of the Andean Montane Forest copes with fragmentation, but may be paying a high cost. *Glob. Ecol. Conserv.* **2022**, *37*, e02174. [CrossRef]
26. Demerdzhiev, D.A.; Horváth, M.; Kovács, A.; Stoychev, S.A.; Karyakin, I.V. Status and population trend of the eastern imperial eagle (*Aquila heliaca*) in Europe in the period 2000–2010. *Acta Zool. Bulg.* **2011**, *3*, 5–14.
27. Ferguson-Lees, J.; Christie, D.A. *Raptors of the World*; Christopher Helm: London, UK, 2001.
28. Abuladze, A. Ecology of the Imperial Eagle *Aquila heliaca* in Georgia. In *Eagle Studies*; Meyburg, B.-U., Chancellor, R.D., Eds.; WWGBP: Berlin, Germany; Paris, France; London, UK, 1996; pp. 447–457.
29. Chavko, J.; Danko, Š.; Obuch, J.; Mihók, J. The Food of the Imperial Eagle (*Aquila heliaca*) in Slovakia. *Slov. Rap. J.* **2007**, *1*, 1–18. [CrossRef]
30. Demerdzhiev, D.; Dobrev, D.; Stoychev, S.; Terziev, N.; Spasov, S.; Boev, Z. Distribution, abundance, breeding parameters, threats and prey preferences of the eastern imperial eagle (*Aquila heliaca*) in European Turkey. *Slovak Raptor J.* **2014**, *8*, 17–25. [CrossRef]
31. Horváth, M.; Solti, B.; Fatér, I.; Juhász, T.; Haraszthy, L.; Szitta, T.; Ballók, Z.; Pásztory-Kovács, S. Temporal changes in the diet composition of the Eastern Imperial Eagle (*Aquila heliaca*) in Hungary. *Ornis Hung.* **2018**, *26*, 1–26. [CrossRef]

32. Karyakin, I.; Kovalenko, A.; Levin, A.; Pazhenkov, A. Eagles of the Aral—Caspian Region, Kazakhstan. *Rap. Con.* **2011**, *22*, 92–152.
33. Karyakin, I.; Nikolenko, E.; Levin, A.; Kovalenko, A. Imperial Eagle in Russia and Kazakhstan: Population status and trends. *Rap. Con.* **2008**, *14*, 18–27.
34. Katzner, T.; Bragin, E.; Knick, S.; Smith, A. Spatial structure in the diet of Eastern Imperial Eagle *Aquila heliaca* in Kazakhstan. *J. Avian Biol.* **2006**, *37*, 594–600. [CrossRef]
35. BirdLife International. Species Factsheet: *Aquila heliaca*. Available online: http://www.birdlife.org (accessed on 24 October 2020).
36. Demerdzhiev, D.; Dobrev, D.; Popgeorgiev, G.; Stoychev, S. Landscape alteration affects the demography of an endangered avian predator by reducing the habitat quality. *Avian Res.* **2022**, *13*, 100030. [CrossRef]
37. Lazarova, I.; Dobrev, D.; Gradev, G.; Petrov, R.; Stoychev, S.; Klisurov, I.; Demerdzhiev, D. Main mortality factors for the Eastern Imperial Eagle (*Aquila heliaca* Savigny, 1809) in Bulgaria. *Orn. Hung.* **2020**, *28*, 120–134. [CrossRef]
38. Hadad, E.; Zduniak, P.; Yosef, R. Sustaining Increasing Wintering Raptor Populations in Central Israel: A 38 Years Perspective. *Sustainability* **2022**, *14*, 12481. [CrossRef]
39. Angelov, I. Numbers, Ecology, Behavior, and Measures for the Conservation of the Imperial Eagle (*Aquila heliaca*) in the Vicinity of the Town of Sliven. Master's thesis, Department of Zoology and Anthropology, Faculty of Biology, Sofia University, Tokyo, Japan, 2009; 82p. (In Bulgarian).
40. Tzonev, R.; Gussev, C.; Popgeorgiev, G. Scrub and grassland habitats of Besaparski Ridove Special Protection Area (Natura 2000), southern Bulgaria: Distribution and assessment of their conservation status. *Acta Zool. Bulg.* **2014**, *5*, 137–142.
41. Forsman, D. Eastern Imperial Eagle Plumages. *Alula* **2005**, *4*, 147–152.
42. Cresswell, W. Flocking is an effective anti-predation strategy in red-shanks, Tringa totanus. *Anim. Behav.* **1994**, *47*, 433–442. [CrossRef]
43. Cresswell, W.; Quinn, J.L. Faced with a choice, sparrowhawks more attack the more vulnerable prey group. *Oikos* **2004**, *104*, 71–76. [CrossRef]
44. Collopy, M.W. Foraging Behavior and Success of Golden Eagles. *Ornithology* **1983**, *100*, 747–749. [CrossRef]
45. Watson, J. *The Golden Eagle*, 2nd ed.; Yale University Press: New Haven, CT, USA, 2010; 374p.
46. Ellis, D.H.; Schmitt, N.J. *Behavior of the Golden Eagle. An Illustrated ethogram*; Hancock House Publishers: Surrey, BC, Canada, 2017; 104p.
47. Gotelli, N.J.; Graves, G.R. *Null Models in Ecology*; Smithsonian Institution Press: Washington, DC, USA, 1996.
48. Shannon, C.E.; Weaver, W. *The Mathematical Theory of Communication*; University of Illinois Press: Urbana, IL, USA, 1949.
49. Chao, A.; Lee, S.M. Estimating the number of classes via sample coverage. *J. Am. Stat. Assoc.* **1992**, *87*, 210–217. [CrossRef]
50. Chao, A.; Shen, T.J. Nonparametric estimation of Shannon's diversity index when there are unseen species in sample. *Environm. Ecolog. Stat.* **2003**, *10*, 429–443. [CrossRef]
51. Burnham, K.P.; Anderson, D.R. *Model Selection and Multimodel Inference: A Practical Information-Theoretic Approach*, 2nd ed.; Springer: New York, NY, USA, 2002; ISBN 0-387-95364-7.
52. StatSoft Inc. STATISTICA (Data Analysis Software System), Version 12. StatSoft Inc.: Tulsa, OK, USA, 2013; Available online: www.statsoft.com (accessed on 9 September 2022).
53. R Core Team, R. *A Language and Environment for Statistical Computing*; R Foundation for Statistical Computing: Vienna, Austria, 2012.
54. Hammer, O.; Harper, D.A.T.; Ryan, P.D. PAST: Paleontological Statistics software package for education and data analysis. *Paleont. Elec.* **2001**, *4*, 9.
55. Heath, J.A.; Kochert, M.N.; Steenhof, K. Golden Eagle dietary shifts following wildfire and shrub loss have negative consequences for nestling survivorship. *Ornithol. Appl.* **2021**, *123*, duab034. [CrossRef]
56. Milchev, B. Diet shifting of tortoise-eating Golden Eagles (*Aquila chrysaetos*) in southeastern Bulgaria. *Ornis Fenn.* **2022**, *99*, 60–70. [CrossRef]
57. Brockmann, H.; Barnard, C. Kleptoparasitism in birds. *Anim. Behav.* **1979**, *27*, 487–514. [CrossRef]
58. Steele, W.; Hockey, P. Factors Influencing Rate and Success of Intraspecific Kleptoparasitism among Kelp Gulls (*Larus dominicanus*). *Ornithology* **1995**, *112*, 847–859. [CrossRef]
59. Cresswell, W. Kleptoparasitism rates and aggressive interactions between raptors. In *Raptors Worldwide: Proceedings of the 6th Conference on Birds of Prey and Owls*; Chancellor, R.D., Meyburg, B.U., Eds.; World Working Group on Birds of Prey and Owls: Budapest, Hungary, 2004; pp. 806–813.
60. Edwards, G.P. Predicting seasonal diet in the yellow-bellied marmot: Success and failure for the linear programming model. *Oecologia* **1997**, *112*, 320–330. [CrossRef] [PubMed]
61. Meyburg, B.U.; Scheller, W.; Bergmanis, U. Home range size, habitat utilization, hunting and time budgets of Lesser Spotted Eagles Aquila pomarina with regard to disturbance and landscape fragmentation. In *Raptors Worldwide*; Chancellor, R.D., Meyburg, B.U., Eds.; WWGBP & MME: Berlin, Germany, 2004.
62. Mirski, P. Effect of selected environmental factors on hunting methods and hunting success in the lesser spotted eagle Aquila pomarina in North-Eastern Poland. *Russ. J. Ecol.* **2010**, *41*, 197–200. [CrossRef]
63. Nadjafzadeh, M.; Hofer, H.; Krone, O. Sit-and-wait for large prey: Foraging strategy and prey choice of White-tailed Eagles. *J. Ornithol.* **2016**, *157*, 165–1178. [CrossRef]

64. Kitowski, I. Age-related differences in foraging behavior of Montagu's harrier *Circus pygargus* males in south-east Poland. *Acta Ethol.* **2003**, *6*, 35–38. [CrossRef]
65. Rutz, C.; Whittingham, M.J.; Newton, I. Age-Dependent Diet Choice in an Avian Top Predator. *Proc. Biol. Sci.* **2019**, *273*, 579–586. Available online: https://www.jstor.org/stable/25223332 (accessed on 12 October 2020). [CrossRef]
66. Marchetti, K.; Price, T. Differences in the Foraging of Juvenile and Adult Birds: The Importance of Developmental Constraints. *Biol. Rev.* **1989**, *64*, 51–70. [CrossRef]
67. Koshev, Y.; Kachamakova, M.; Arangelov, S.; Ragyov, D. Translocations of European ground squirrel (*Spermophilus citellus*) along altitudinal gradient in Bulgaria—An overview. *Nat. Conserv.* **2006**, *35*, 63–95. [CrossRef]
68. Demerdzhiev, D. Breeding parameters and factors infuencing the reproduction of an expanding Long-legged Buzzard (Buteo rufnus) population under high breeding density conditions. *J. Ornithol.* **2022**, *163*, 405–415. [CrossRef]
69. Mañosa, S.; Real, J.; Codina, J. Selection of settlement areas by juvenile Bonelli's eagle in Catalonia. *J. Raptor. Res.* **1988**, *32*, 208–214.
70. Kachamakova, M.; Koshev, Y.; Rammou, D.-L.; Spasov, S. Rise and fall: Results of a multidisciplinary study and 5-year long monitoring of conservation translocation of the European ground squirrel. *Biodivers. Data J.* **2022**, *10*, e83321. [CrossRef]
71. Demerdzhiev, D. Birds in Besaparski riodve special protection area (Natura 2000), Southern Bulgaria: Conservation status and dynamics. *Acta Zool. Bulg. Suppl.* **2014**, *5*, 171–189.

diversity MDPI

Article

Does Temporal and Spatial Diet Alteration Lead to Successful Adaptation of the Eastern Imperial Eagle, a Top Predator?

Dimitar Demerdzhiev [1,2,*], Zlatozar Boev [2], Dobromir Dobrev [1], Nedko Nedyalkov [2] and Tseno Petrov [1]

1 Bulgarian Society for the Protection of Birds, Nikola Vaptsarov Str. 3, Asenovgrad/BirdLife Bulgaria, Leonardo Da Vinci Str. 5, 4000 Plovdiv, Bulgaria
2 National Museum of Natural History, Bulgarian Academy of Sciences, Blvd. Tsar Osvoboditel 1, 1000 Sofia, Bulgaria
* Correspondence: dimitar.demerdzhiev@bspb.org

Abstract: Predator–prey interactions may be linked to different temporal or spatial patterns, including dynamics in prey populations. Therefore, understanding the adaptive capacity and how top predators respond to shifts in prey abundance and availability is crucial for their conservation. In this study, we investigated the diet pattern of the endangered Eastern Imperial Eagle facing long-term and large-scale changes. We studied the abundance variation of its profitable prey, sousliks, and how it reflected on eagle population trajectories in a regional and temporal context. We found a significant diet alteration expressed in large decrease of brown hare ($\beta^2 = -0.83$), poultry ($\beta^2 = -0.81$), gulls ($\beta^2 = -0.71$), and water birds ($\beta^2 = -0.57$), and an obvious increase of northern white-breasted hedgehog ($\beta^2 = 0.61$) and doves ($\beta^2 = 0.60$). Raptors and owls raised their participation ($\beta^2 = 0.44$), but white stork and different reptiles supplied more biomass. Abundance of European souslik decreased through the studied periods (adjusted $R^2 = 0.25$, $p < 0.001$) which accounted for the lower proportion of this prey in the eagle's diet. Nevertheless, the eagle population successfully adapted and significantly increased ($\beta^2 = 0.97$) in most of the distribution area. The trophic strategy used by this top predator related to opportunistic foraging represents an ecological advantage that allows the species to adapt to different habitats and guarantees its future. The observed prolonged diet alteration could result in a significant negative attitude among different groups such as hunters, pigeon fanciers, and poultry keepers towards eagles. Therefore, enhanced communication with key stakeholders is needed. Conservation efforts should be focused also on the preservation of the species' main foraging habitats and the restoration of damaged ones so as to maintain the good conditions of both primary food source and subsequent prey.

Keywords: *Aquila heliaca*; food spectrum; diet changes; diurnal raptors; long-term studies; generalist; adaptation; prey

Citation: Demerdzhiev, D.; Boev, Z.; Dobrev, D.; Nedyalkov, N.; Petrov, T. Does Temporal and Spatial Diet Alteration Lead to Successful Adaptation of the Eastern Imperial Eagle, a Top Predator? *Diversity* **2022**, *14*, 1000. https://doi.org/10.3390/d14111000

Academic Editor: Michael Wink

Received: 11 October 2022
Accepted: 16 November 2022
Published: 19 November 2022

1. Introduction

Large raptor species are limited by different factors such as food supply, nest-site availability, weather conditions, and bird experience [1–3]. Generalist predators can change their diet mainly in response to habitat alteration and depletion of main food resources. Such changes in diet can affect population trajectories via individual fitness and breeding performance [4,5]. Therefore, understanding the adaptive capacity and how top predators respond to shifts in prey abundance and availability is crucial for their conservation. Successful adaptation of top predators to changes in availability and abundance of main prey sources determines their ability to survive and expand their populations in a changing environment.

It is generally considered that generalists are more adaptable to spatially or temporally heterogeneous environments, while specialists are more adaptable to temporally stable environments [6]. Since generalist species may have wider dietary niches and can switch between prey resources, when the preferred prey declines spatially or temporally, they are less susceptible to the negative demographic effect caused by changes in prey availability and abundance than more specialized species [7], but see 39. However, the classification of generalist or specialist can occur along a gradient of adaptability and, furthermore, a generalist species can be made up of specialized individuals ([8], as well as a typical "specialist" can successfully adopt a generalist foraging strategy [9]. Then, within a species range, the individual's capacity to utilize alternative resources is crucial for successful adaptation when the main prey is depleted. In fact, the response at the individual level may vary depending on how individuals rank their prey, which, in turn, results in different resource use patterns [10–12].

Processes in ecology vary over time [13], and predator–prey interactions may be linked to different temporal or spatial patterns, including cycles and outbreaks in prey populations [14,15], leading to spatial and temporal shifts in the predator's diet [16,17]. Therefore, in this predator–prey system, the predator may include new alternative prey sources when preferred prey is scarce [18,19].

Here, we examine a generalist predator, the eastern imperial eagle (*Aquila heliaca*), hereafter EIE, foraging in open habitats with predominant grass vegetation [20], where it exploits various prey species of different size [21–26]. Different souslik species (*Spermophilus* sp.) represent profitable prey of this eagle, determining the distribution and density of the largest Eastern populations in Russia and Kazakhstan [26–28]. However, in other parts of the species' distribution area, the lack of large continuous souslik colonies leads to dietary shifts and wider prey diversity [21–25,29]. Previous studies recorded regional diet differences in terms of subpopulations [24,26], and those differences were strongly influenced by the individual territories occupied by the eagles [22]. Temporal changes in the EIE diet are well-documented only for the westernmost Pannonian population, where traditional prey species such as common hamster (*Cricetus cricetus*) and European souslik (*Spermophilus citellus*) are shifted by corvids (*Corvidae*), water birds, and roe deer (*Capreolus capreolus*) [24].

In this study, we investigated if the diet pattern of an endangered top predator such as the EIE faced long-term and large-scale changes, and if so, how the eagle responded to such shifts. We studied the abundance variation of its profitable prey, such as sousliks, and how it reflected on the EIE population trajectories in a regional and temporal context.

Our aim was to explore the adaptive capacity of this generalist species describing the mechanism of changes in the resource use pattern by which it switched between the different food sources.

We predicted that if the availability and abundance of profitable prey decreased, eagles could substitute the decreasing prey with other plentiful food sources and thus survive and increase their number, and vice versa—the lack of sufficiently abundant and accessible prey would lead to territory abandonment and population decline. We studied which species and to what extent could substitute the decreasing prey and planned future conservation strategies.

2. Materials and Methods

2.1. Study Area

The diet remains were collected in the whole distribution area of the species in Bulgaria [22]. We sampled 37 different breeding territories, distributed among six geographical units (Figure 1). Mountainous habitats (ER, SG) were characterized by small fragments of pastures and meadows and considerable forest cover, where eagles bred close to the forest edge, using common beech (*Fagus sylvatica*), sessile oak (*Quercus petraea*), and scots pine (*Pinus silvestris*) [30]. We merged territories from SG (n = 2) and ER (n = 1) into a group of high mountain regions (HM) due to the small sample sizes and the similar habitat conditions [22]. The EIEs in other regions occupied hilly areas and lowlands, where grasslands, usually overgrown with shrub formations of oriental hornbeam (*Carpinus orientalis*) and Christ's thorn (*Paliurus spina-christi*), agricultural fields and small forest patches formed a mosaic habitat structure. They built their nests on single trees or in small groups of trees, mainly hybrid poplars (*Populus* sp.) or Hungarian oak (*Quercus frainetto*), downy oak (*Quercus pubescens*), and Turkey oak (*Quercus cerris*), often along small streams or in fields.

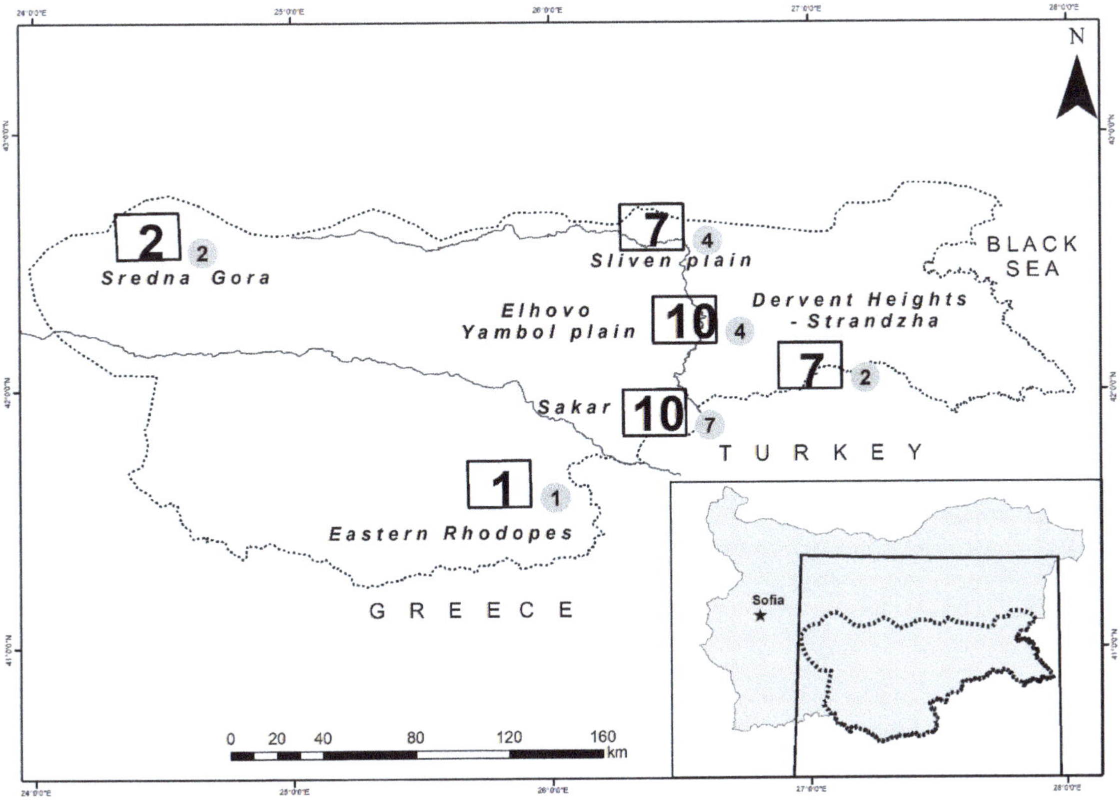

Figure 1. Number of sampled breeding territories of the eastern imperial eagle (in a square) vs. number of sampled souslik plots (in circle) in the different regions (Eastern Rhodope Mnt., ER; Sredna Gora Mnt., SG; Sliven plain, SP; Elhovo-Yambol plain, EYP; Dervent Heights-Western foothills of Strandzha Mnt., DHWstr; Sakar Mnt.).

2.2. Data Collection

In this study, we used a 23 years' data set (1999–2021), part of which has already been published, although in a different context [22]. The first detailed study on the diet of the EIE from Bulgaria considered only food composition, regional distribution, and seasonal differences, not reporting the temporal changes [22]. In this study, we analyzed the temporal variation in main prey species, using data about 5315 prey specimens, covering the entire distribution area of the EIE in the country (Table 1). The annual number of sampled EIE territories corresponded to the number of occupied territories ($r_c = 0.81$, $p = 0.0001$) (Figure 2).

Each nesting site was visited twice in each of the following periods: November–February, June–August (post-fledging period). Food remains, bones, feathers, and pellets were collected inside and under nests and roosts [31]. The following types of remains were not included in the data in order to reduce the bias of indirect sampling, even if they were found under the nest sites or roosting trees: (1) single feathers, which could be shed by live birds; (2) full carcasses of large animals, which could not be brought there by the eagles; (3) old or deteriorated samples, which could have remained from previous years [22,24]. The material was identified through the comparative osteological collections of the National

Museum of Natural History at the Bulgarian Academy of Sciences. Whenever possible, the minimum number of individuals (MNI) in each pellet or prey remain was estimated based on the number of skeletal or keratinized body parts [22,31]. The MNI was determined by taking into account the age (juvenis, subadultus, adultus), sex, and the size differences between individuals.

Table 1. Number of sampled prey specimens and number of sampled eastern imperial eagle occupied territories in different regions.

Regions	Number of Sampled Occupied Territories	Number of Sampled Prey Specimen
HM (Sredna Gora Mnt. and Eastern Rhodope Mnt.)	3	456
Sakar Mnt.	10	2090
DHWstr (Dervent Heights-Western foothills of Strandzha Mnt.)	7	1239
EYP (Elhovo-Yambol plain)	10	916
SP (Sliven plain)	7	614
TOTAL	**37**	**5315**

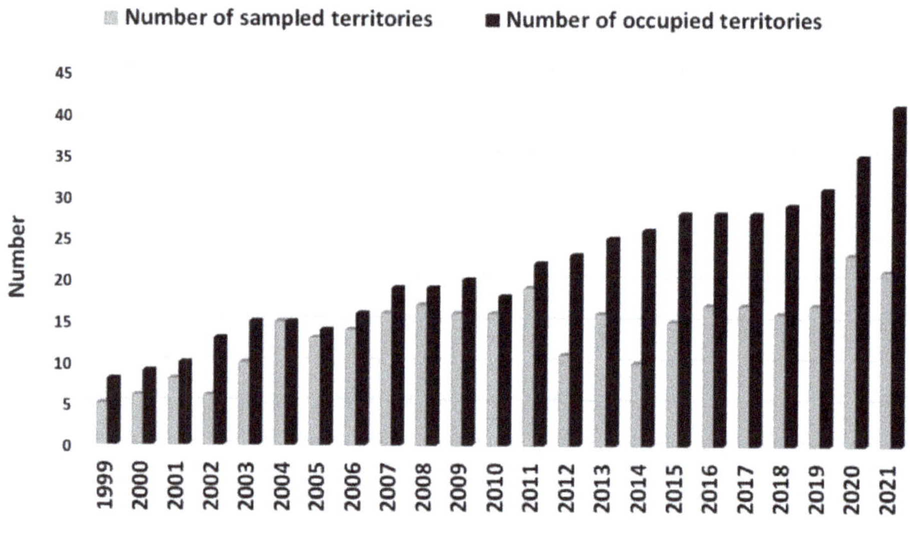

Figure 2. Number of sampled eastern imperial eagle territories and number of occupied territories in different years.

The body mass of the specimens of the various species was determined by [32–39]. An average body mass was given, calculated on the basis of the average mass of individual specimens. When the material was identified up to genus level, the average values for the presented species of the genus were given. The carrion biomass was not taken into account [22].

2.3. *Profitable Prey Abundance*

Prey abundance of profitable prey (European souslik) was estimated through test plots, each covering 1 ha, where all active holes were counted and recorded [40]. In total, 20 such plots were monitored during the whole study. The plots were located in the souslik colonies distributed in all the studied regions within the EIE's occupied territories (Figure 1). To avoid the effect of cycles in rodent abundance between years, each plot was visited in two consecutive years of the three study periods (see below). In peak years, rodents would be highly abundant, while in the poor years, the opposite will be valid [3]. The count of the abundance of sousliks was carried out twice in the studied year, in the months of April and May. The mean value of the reported individuals (active holes) per plot in a given year was taken in the analysis.

2.4. Data Analyses

In order to identify the main changes in the diet composition, the prey items were grouped into the main categories, following the already published methodology [22]: lizards and snakes (Squamata), tortoises (Testudines), water birds (Anatidae, Ardeidae), poultry (*Gallus gallus f. domestica*, *Anser anser f. domestica*, *Meleagris gallopavo f. domestica*, *Pavo cristatus f. domestica*), phasianids (Phasianidae), gulls (Laridae), doves (Columbidae, Feral Pigeon), songbirds (Non-Corvidae Passerines), corvids (Corvidae), stork (*Ciconia ciconia*), raptors and owls (Accipitridae, Falconidae, Strigidae, Tytonidae), hedgehog (*Erinaceus roumanicus*), hare (*Lepus europaeus*), souslik (*Spermophilus citellus*), rodents (Rodentia excl. European souslik), carnivores (Carnivora), carrion (Artiodactyla, Perissodactyla), and other animals (*including other vertebrate taxa*).

We divided our data set into three periods associated with significant changes in the country's land use pattern that might have affected the populations of eagle's prey [41]. The first period (1999–2006) included the years prior to Bulgaria's accession to the European Union (EU). This period was characterized by extensive agriculture and animal husbandry. The second period (2007–2013), related to the country's accession to the EU, was characterized by gradual intensification of some aspects of agriculture through subsidies. During this period, intensive plowing of natural and semi-natural grasslands and their conversion back to arable land was registered [41]. The third period (2014–2021) included the last programming period of the European Commission's CAP (Common Agriculture Policy). It was characterized by a new habitat alteration financially stimulated by subsidies and expressed in large-scale removal of shrubs from grasslands using mechanized equipment such as shredders and bulldozers. We compared the frequency and biomass contribution of the different prey categories among the geographical units and in general during the three periods in order to investigate if there were any evident long-term and large-scale alterations in the diet composition.

As a first step, we applied the over-parameterized linear model (GLM) with Type III error distribution. We ran two models: one for the proportion and one for the biomass, including the year as a continuous covariate and each prey category as explanatory factors. Secondly, we built a simple mixed model (GLMM) including the prey category and the study period as explanatory factors. To control for spatial variation in prey abundance and composition, as well as for possible differences in feeding strategies among eagle populations living in different environments, we included a random factor "region" to account for data pooled within each region. We ran two models again: one for the proportion and one for the biomass. After that we applied a post-hoc analysis (Tukey's HSD test) to extract the significance of the trends of each category.

Changes in profitable prey abundance (souslik density) were evaluated through GLMM where souslik density and study periods were explanatory variables and "region" was included as a random effect. A *post hoc* analysis (Tukey's HSD test) was used to extract the significance of the trends of souslik density in each period from the model.

The design we used could not account for the fact that samples collected in the same nest or nearby trees could be predated in the different years by the same individuals [24]. Eagles breeding in a given nesting site could change over the years, then remains collected in the same nesting sites could derive from independent individuals. Similarly, the remains in a nest in a particular year included items predated by the male or the female of the given pair in an unknown proportion; therefore, the data about the two individuals could not be separated in the individual samples. Hence, the factor "individual eagle" could not be included in our model. Nonetheless, we considered our aim to detect long-term and large-scale changes in the eagles' diet achieved because of the large applicable and representative data sampling.

The data calculated in percentage (prey frequency and biomass contribution) were converted into proportions and then Arcsin transformed to achieve a close to normal distribution [42]. To evaluate the results of the regression models, we used the adjusted R^2 value as a correction factor. We also used explanatory parameter estimates (β^2) with lower and upper CL (95%) and a probability value (p) of the explanatory factors. Results with $p \leq 0.05$ were considered significant. Values were provided as means $\pm$ standard error (SE). All data were analyzed using Statistica for Windows, Release 12 [43].

3. Results

3.1. General Pattern of Main Prey Contribution and EIE Population

The 23 years' trend of the different prey categories showed a significant EIE diet alteration (Table 2). While the share of hare, poultry, and gulls showed the largest decrease, both in terms of frequency and biomass, hedgehogs and doves increased significantly their presence and biomass contribution (Table 2, Figure 3). To a lesser extent, water birds also reduced their occurrence (adjusted $R^2 = 0.29$, $\beta^2 = -0.57$, $p = 0.004$) and biomass supply (adjusted $R^2 = 0.14$, $\beta^2 = -0.42$, $p = 0.047$). Decline was found also for the presence of carrion (adjusted $R^2 = 0.35$, $\beta^2 = -0.61$, $p = 0.002$) (Figure 3). In contrast, raptors and owls rose their participation in the eagle's diet (adjusted $R^2 = 0.16$, $\beta^2 = 0.44$, $p = 0.03$). The share of categories storks and tortoises (adjusted $R^2 = 0.20$, $\beta^2 = 0.49$, $p = 0.02$) also increased their importance to biomass provision, while lizards and snakes had only a marginal effect through the years (adjusted $R^2 = 0.14$, $\beta^2 = 0.42$, $p = 0.046$). However, other animals also increased their biomass supply (adjusted $R^2 = 0.30$, $\beta^2 = 0.58$, $p = 0.004$).

The strongest negative correlation was found between the categories souslik vs. carnivores ($r_c = -0.61$, $p = 0.002$). Of the other prey categories that had demonstrated a significant trend over the years, hedgehog negatively corelated with water birds ($r_c = -0.50$, $p = 0.02$), carnivores ($r_c = -0.47$, $p = 0.02$), and poultry ($r_c = -0.43$, $p = 0.04$). Increasing stork was related with the depression of poultry ($r_c = -0.49$, $p = 0.02$), carrion ($r_c = -0.48$, $p = 0.02$), and hare ($r_c = -0.45$, $p = 0.03$). The decline of hare also corelated with rise of raptors and owls ($r_c = -0.45$, $p = 0.03$), and that of gulls—with the increasing share of doves ($r_c = -0.47$, $p = 0.02$) and raptors and owls ($r_c = -0.43$, $p = 0.04$). However, another significant negative correlation was found between the categories tortoises and songbirds ($r_c = -0.46$, $p = 0.03$).

The EIE population significantly increased between 1999 and 2021 (adjusted $R^2 = 0.95$, $\beta^2 = 0.97$, $p < 0.001$), starting from eight occupied territories and reaching forty-one in the last year of the study (Figure 2).

3.2. Temporal and Spatial Comparison of Eagle Abundance, Profitable Prey Abundance, and Diet Composition in the Studied Periods

Among the three studied periods, the EIE population gradually increased in Sakar Mnt. ($\beta^2 = 0.66$, $p < 0.001$), DHWstr ($\beta^2 = 0.32$, $p < 0.001$), and EYP ($\beta^2 = 0.26$, $p = 0.001$) (Table 3, Figure 4). In contrast, eagle abundance shrank in HM, a process that started in the second period (Tukey's HSD test = 0.045) and was clearly evident in the last one (Tukey's HSD test < 0.001) (Figure 4). In SP, the first pair of EIE occupied the territory in 2007, reaching the maximum number in 2021 (n = 8) (Figure 4). However, in Sakar Mnt., we recorded a significant increase in eagle pairs in the second period (Tukey's HSD test = 0.02), followed by stable population numbers (Tukey's HSD test = 0.90).

Table 2. Results of the over-parameterized linear model (GLM) carried out to analyze the trend of the different prey categories (frequency and biomass contribution) of the eastern imperial eagle between 1999 and 2021. We used adjusted R^2 value as a correction factor, explanatory parameter estimates (β^2) with lower (95%) and upper CL (95%), and a probability value (*p*) of the explanatory factors. Significant values are given in bold.

Prey Categories	Frequency					Biomass				
	Adjusted R^2	$F_{(1,21)}$	β^2	LCL/UCL	*p*	Adjusted R^2	$F_{(1,21)}$	β^2	LCL/UCL	*p*
Lizards and snakes	0.05	2.05	0.30	−0.14/0.73	0.17	**0.14**	**4.51**	**0.42**	**0.009/0.83**	**0.046**
Tortoises	0.12	4.11	0.40	−0.01/0.82	0.055	**0.20**	**6.62**	**0.49**	**0.09/0.89**	**0.02**
Water birds	**0.29**	**10.16**	**−0.57**	**−0.94/−0.20**	**0.004**	**0.14**	**4.46**	**−0.42**	**−0.83/−0.006**	**0.047**
Poultry	**0.64**	**40.19**	**−0.81**	**−1.08/−0.54**	**<0.001**	**0.65**	**42.36**	**−0.82**	**−1.08/−0.56**	**<0.001**
Phasianids	0.01	0.70	0.18	−0.27/0.63	0.41	0.01	1.16	0.23	−0.21/0.67	0.29
Gulls	**0.48**	**21.57**	**−0.71**	**−1.03/−0.39**	**<0.001**	**0.44**	**18.08**	**−0.68**	**−1.01/−0.35**	**<0.001**
Doves	**0.33**	**11.91**	**0.60**	**0.24/0.96**	**0.002**	**0.42**	**16.61**	**0.66**	**0.33/1.00**	**<0.001**
Songbirds	0.05	0.03	−0.04	−0.49/0.41	0.86	0.05	0.02	0.03	−0.42/0.48	0.89
Corvids	0.02	0.51	0.15	−0.29/0.60	0.48	0.04	0.18	0.09	−0.36/0.54	0.68
Stork	0.09	3.23	0.36	−0.06/0.79	0.09	**0.20**	**6.48**	**0.49**	**0.09/0.88**	**0.02**
Raptors and owls	**0.16**	**5.13**	**0.44**	**0.04/0.85**	**0.03**	0.07	2.73	0.34	−0.09/0.77	0.11
Hedgehog	**0.34**	**12.41**	**0.61**	**0.25/0.97**	**0.002**	**0.50**	**22.95**	**0.72**	**0.41/1.04**	**<0.001**
Hare	**0.67**	**44.79**	**−0.83**	**−1.08/−0.57**	**<0.001**	**0.67**	**46.19**	**−0.83**	**−1.08/−0.58**	**<0.001**
Souslik	0.03	0.41	−0.14	−0.59/0.31	0.53	0.04	0.20	−0.10	−0.55/0.35	0.66
Rodentia (excluding souslik)	0.03	0.45	−0.14	−0.59/0.30	0.51	0.05	0.01	−0.005	−0.46/0.45	0.98
Carnivores	0.01	1.16	−0.23	−0.67/0.21	0.29	0.05	0.01	0.02	−0.43/0.47	0.92
Carrion	**0.35**	**12.70**	**−0.61**	**−0.97/−0.26**	**0.002**	NA	NA	NA	NA	NA
Other animals	0.11	3.64	0.38	−0.035/0.80	0.07	**0.30**	**10.47**	**0.58**	**0.21/0.95**	**0.004**

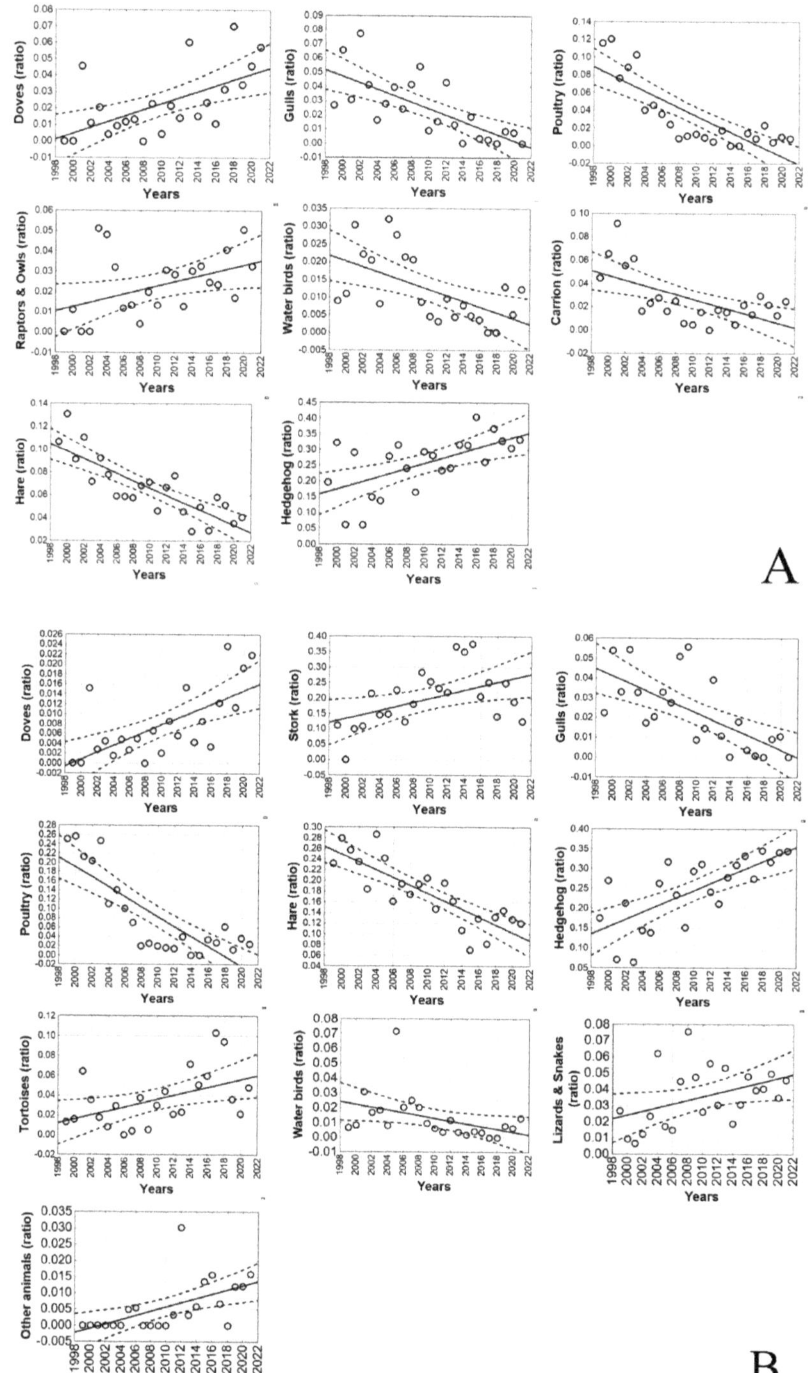

Figure 3. Results of the general linear model (GLM) showing a significant trend in frequency (**A**) and biomass (**B**) of different prey categories between 1999 and 2021.

Table 3. Results of the general linear mixed model (GLMM) carried out to analyze the trend of souslik abundance and eastern imperial eagle (EIE) abundance in different regions. We used explanatory parameter estimates (β^2) ± standard error, with lower (95%) and upper CL (95%) and a probability value (p) of the tested categories. Significant values are given in bold.

Categories	Effect	Region	β^2	Std. Err	LCL/UCL	t	p
Souslik abundance	**Random**	**Sakar Mnt.**	**0.46**	**0.20**	**−0.07/0.86**	**2.38**	**0.02**
	Random	**SP**	**0.38**	**0.18**	**−0.01/0.74**	**2.09**	**0.04**
	Random	EYP	0.30	0.18	−0.06/0.66	1.66	0.10
	Random	HM	0.24	0.16	−0.07/0.56	1.55	0.13
	Random	DHWstr	no tolerance	no tolerance	no tolerance	no tolerance	no tolerance
EIE abundance	**Random**	**Sakar Mnt.**	**0.66**	**0.08**	**0.51/0.81**	**8.73**	**<0.001**
	Random	SP	no tolerance	no tolerance	no tolerance	no tolerance	no tolerance
	Random	**EYP**	**0.26**	**0.08**	**0.10/0.41**	**3.36**	**0.001**
	Random	HM	−0.11	0.08	−0.26/0.04	−1.44	0.15
	Random	**DHWstr**	**0.32**	**0.08**	**0.17/0.47**	**4.23**	**<0.001**

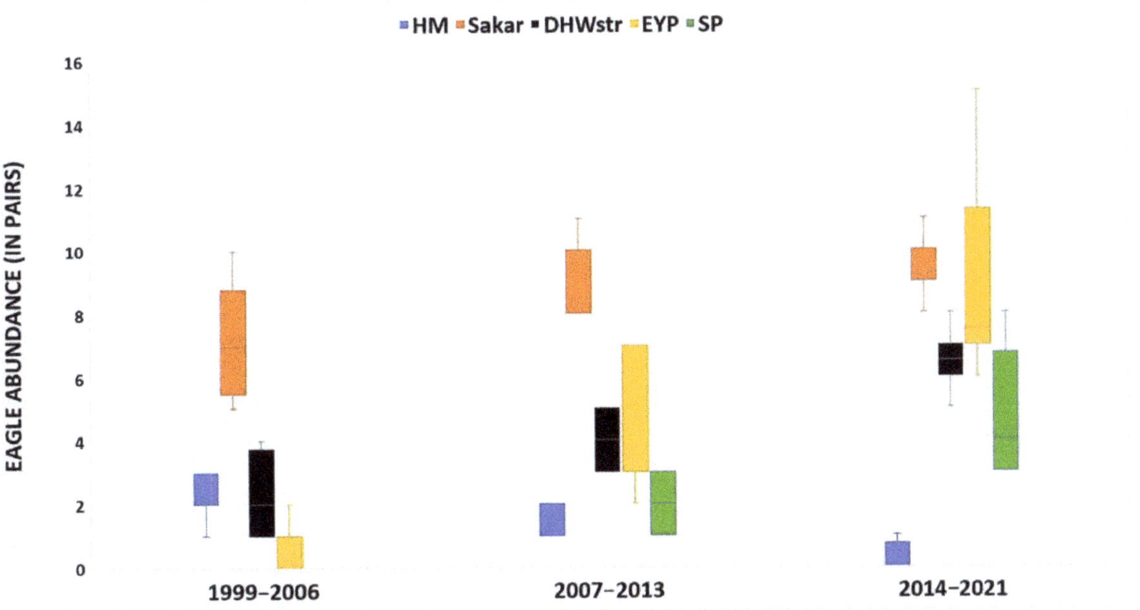

Figure 4. Spatial dynamics of the eastern imperial eagle population (Eastern Rhodope Mnt. and Sredna Gora Mnt., HM; Sliven plain, SP; Elhovo-Yambol plain, EYP; Dervent Heights-Western foothills of Strandzha Mnt., DHWstr; Sakar Mnt.).

Abundance of profitable prey sousliks decreased over the study (adjusted $R^2 = 0.25$, $F_2 = 8.97$, $p < 0.001$), a process clearly evident in the third period (Tukey's HSD test = 0.002). A gradual decline was reported in Sakar Mnt. ($\beta^2 = 0.46$, $p = 0.02$) and SP ($\beta^2 = 0.38$, $p = 0.04$), while species' abundance in DHWstr did not show any trend (Table 3, Figure 5). However, souslik abundance in EYP and high mountains (SG and ER) slightly increased in the second period, followed by a severe drop in the last one (Figure 5).

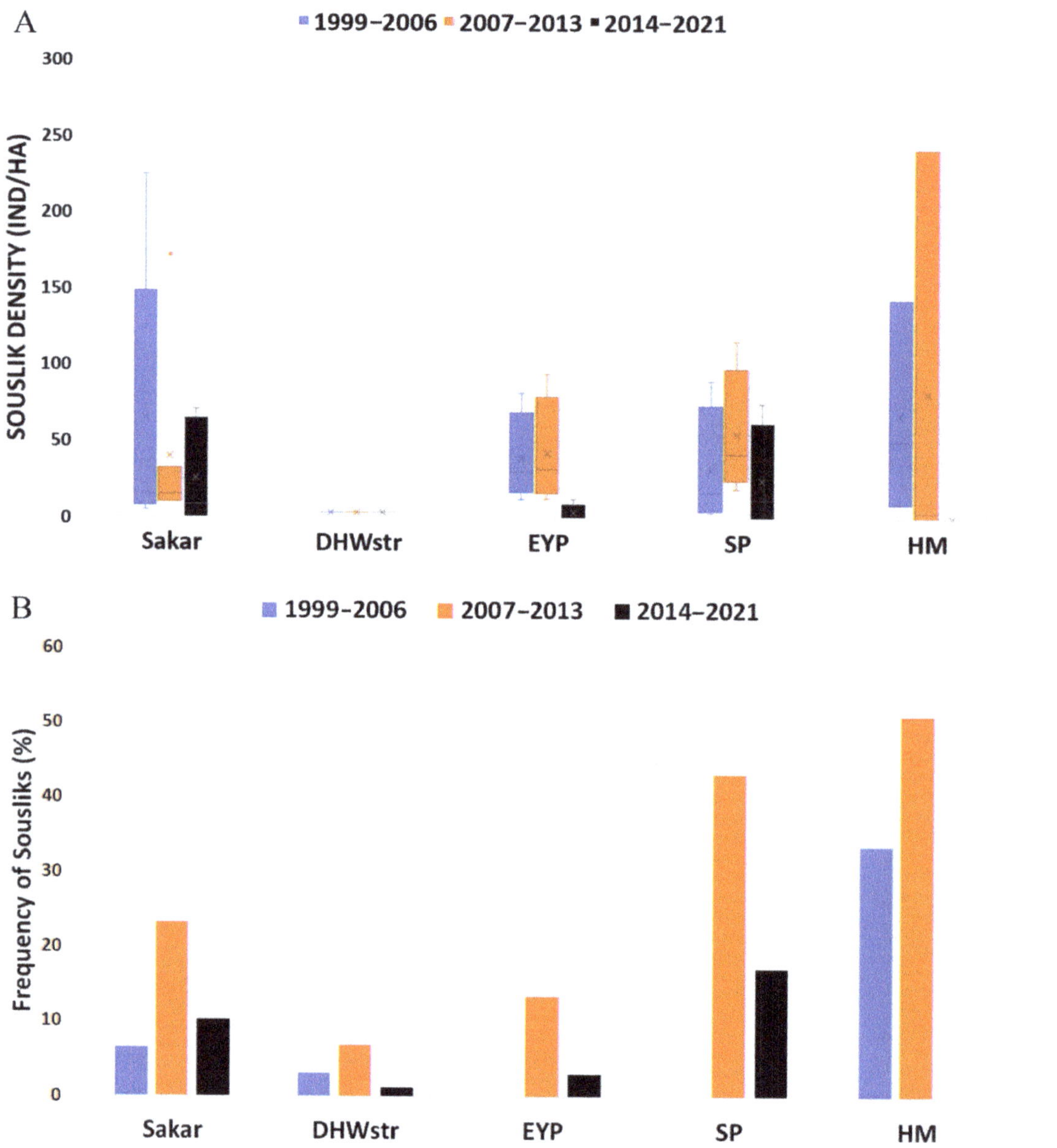

Figure 5. Spatial dynamics of souslik abundance (ind/ha) (**A**) and frequency in the eagle's diet (%) (**B**) (Eastern Rhodope Mnt. and Sredna Gora Mnt., HM; Sliven plain, SP; Elhovo-Yambol plain, EYP; Dervent Heights-Western foothills of Strandzha Mnt., DHWstr; Sakar Mnt.).

The frequency of sousliks in the EIE's diet followed more or less the same pattern as that of souslik abundance (Figure 5). In the second period, the presence of this profitable prey marginally increased (Tukey's HSD test = 0.053), followed by a dramatic decline in the last one (Tukey's HSD test = 0.008). In Sakar Mnt., sousliks reached 23.3% of caught prey between 2007 and 2013 and then dropped to 10.21% ($\beta^2 = -0.73$, $p = 0.004$). This prey also shrank significantly its presence in EYP ($\beta^2 = -0.80$, $p = 0.003$). With regard to the decline of souslik abundance in SP, their frequency was reduced from 43.20% in the second period to 17.16% in the last one. Following the dynamics of the main prey, in high mountains the proportion of sousliks increased in the second period, accounting for 51% of the consumed prey, and then practically disappeared from the eagle's diet in the third one.

Comparing the three studied periods, two prey categories showed obvious significant differences, both with regard to presence and biomass contribution (Table 4). In the second period, lizards and snakes increased their proportion (Tukey's HSD test = 0.05) and biomass participation (Tukey's HSD test = 0.006). This increment was obvious in Sakar Mnt. ($\beta^2 = 1.11$, $p < 0.001$) and DHWstr ($\beta^2 = 0.61$, $p = 0.006$), where this category reached 9.52% and 7.74% of the caught prey, respectively. The other category, water birds, decreased severely in all of the studied regions (Table S1).

One of the most important prey categories, hedgehog, increased its proportion, a process clearly visible in the last period (Tukey's HSD test = 0.03) (Table 4). This phenomenon was observed mostly in Sakar Mnt. in terms of proportion ($\beta^2 = 0.66$, $p = 0.02$) and biomass participation ($\beta^2 = 0.66$, $p = 0.046$). Similarly, in DHWstr, the ratio of hedgehogs in the eagle's diet rose notably ($\beta^2 = 0.70$, $p = 0.01$), starting from 20.54% in the first period and reaching 50.30% in the last one. However, biomass supply by hedgehogs also grew in value from 15.30% to 40.11%, although this change was not statistically significant ($\beta^2 = 0.58$, $p = 0.068$).

Comparing the three studied periods, brown hare (*Lepus europaeus*) also diminished its importance in the EIE's diet (Table 4), a process showing a significant trend in EYP ($\beta^2 = 1.05$, $p = 0.002$), where biomass provided by this prey dropped from 39.08% to 14.90%. Another prey category, rodents, significant declined in the eagle's diet in Sakar Mnt. and DHWstr ($\beta^2 = -0.93$, $p = 0.01$), despite the fact that its meaning to biomass was marginal. However, in HM, the frequency of rodents in the eagle's diet dropped from 49.03% to 18%.

Between the periods, the category stork increased its share (adjusted $R^2 = 0.69$, $F_6 = 5.00$, $p = 0.049$) and biomass participation (adjusted $R^2 = 0.68$, $F_6 = 4.88$, $p = 0.052$) in the EIE's diet (Table 4), but in regional context we observed a significant decline, both in presence ($\beta^2 = 0.99$, $p = 0.01$) and biomass contribution ($\beta^2 = 0.84$, $p = 0.02$), only in DHWstr. In the third period, the eagles breeding in this region reduced the presence of storks in their menu from 17.57% to 10.54%, while biomass provided from storks shrank from 41.9% to 27.2%.

The contribution of songbirds demonstrated different patterns in different regions. They significantly declined in the eagle's diet in Sakar Mnt. ($\beta^2 = 0.78$, $p = 0.004$) and increased in EYP ($\beta^2 = 0.82$, $p = 0.003$) and SP ($\beta^2 = 1.18$, $p < 0.001$). Four other bird prey categories showed regional trends (Table S1). Raptors and owls increased their proportion ($\beta^2 = 1.00$, $p = 0.04$) and biomass supply ($\beta^2 = 1.09$, $p = 0.02$) in EYP, but corvids reduced their share in SP from 9.71% to 5.15% ($\beta^2 = 1.08$, $p = 0.02$), and gulls in DHWstr from 7.95% to 1.39% ($\beta^2 = 0.95$, $p = 0.04$). Phasianids' frequency rose in SP ($\beta^2 = 0.99$, $p = 0.01$) and dropped in EYP ($\beta^2 = 1.15$, $p = 0.005$), although their importance was negligible in general (Table 4). However, in EYP, other vertebrate preys decreased their frequency from 10.58% to 3.81% ($\beta^2 = 0.90$, $p = 0.03$) and their biomass contribution ($\beta^2 = 1.03$, $p < 0.001$).

Table 4. Proportion of different eastern imperial eagle prey categories in Bulgaria in the three studied periods.

Prey Categories	1999–2006			2007–2013			2014–2021			Total (1999–2021)		
	Number	% N	% Biomass	Number	% N	% Biomass	Number	% N	% Biomass	Number	% N	% Biomass
Lizards and snakes	63	4.88	2.47	154	7.82	4.83	133	6.47	3.96	350	6.59	3.91
Tortoises	17	1.32	2.04	28	1.42	2.20	73	3.55	6.01	118	2.22	3.64
Water birds	24	1.86	2.25	21	1.07	1.16	11	0.54	0.47	56	1.05	1.16
Poultry	77	5.97	16.53	26	1.32	3.10	18	0.88	2.52	121	2.28	6.22
Phasianids	24	1.86	0.87	30	1.52	0.80	47	2.29	1.12	101	1.90	0.94
Gulls	42	3.26	2.98	57	2.89	2.99	11	0.54	0.55	110	2.07	2.04
Doves	12	0.93	0.29	38	1.93	0.65	74	3.60	1.30	124	2.33	0.81
Songbirds	50	3.88	0.30	70	3.55	0.23	75	3.65	0.30	195	3.67	0.27
Corvids	36	2.79	1.30	64	3.25	1.50	83	4.04	1.76	183	3.44	1.55
Stork	59	4.57	13.88	143	7.26	23.82	144	7.01	23.21	346	6.51	21.11
Raptors and owls	28	2.17	1.43	35	1.78	1.14	66	3.21	1.69	129	2.43	1.43
Hedgehog	237	18.37	18.46	493	25.03	24.88	653	31.78	31.72	1383	26.02	25.93
Hare	103	7.98	22.38	123	6.24	18.03	83	4.04	11.28	309	5.81	16.50
Souslik	146	11.32	3.90	431	21.88	7.92	141	6.86	2.44	718	13.51	4.80
Rodents	262	20.31	1.07	147	7.46	0.60	295	14.36	0.99	704	13.25	0.87
Carnivores	55	4.26	9.58	45	2.28	5.64	69	3.36	9.56	169	3.18	8.14
Carrion	45	3.49	NA	24	1.22	NA	34	1.65	NA	103	1.94	NA
Other animals	10	0.77	0.28	41	2.08	0.50	45	2.19	1.13	96	1.81	0.69
Total	**1290**	**100.00**	**100.00**	**1970**	**100.00**	**100.00**	**2055**	**100.00**	**100.00**	**5315**	**100.00**	**100.00**

4. Discussion

4.1. Long-Term and Large-Scale Changes in the Diet Pattern

Our results clearly demonstrated the prolonged and wide-reaching diet alteration pattern of a generalist top predator. We found that the previously important brown hare and poultry became of less importance, while the northern white-breasted hedgehog (*Erinaceus roumanicus*), white stork (*Ciconia ciconia*), and doves remarkably increased their significance. The ratio of gulls, water birds, and carrion showed a notable decrease, although their roles were marginal. In parallel with the loss of those preys, the categories raptors and owls, lizards and snakes, and tortoises became regularly detected.

In the beginning of the 21st century, poultry and brown hare were found to be the main prey of the EIE in the southeastern part of the country [44]. At the same time, poultry was mentioned as a primary food source for only one pair in ER [30]. We assume that the abandonment of poultry as a food source by eagles was due to a change in bird farming practices and the demographic decline of the population in Bulgaria. Following the disintegration of the communist regime in the country, there was a clear trend of migration of the population to the major cities. Thus few, mostly elderly people, who no longer kept livestock, remained in the small settlements (near which the eagles' nests were located). In addition, a strict order introduced after 2006, related to cases of avian influenza and banning of free-range poultry farming, severely limited the possibilities for eagles to catch such prey.

The decline of brown hare in the EIE's diet corresponded to the reported decrease of the species' population in Bulgaria in the past decades, especially for EYP [45]. The significant transformation of grasslands [41] expressed in the removal of shrub vegetation through shredders and bulldozers shrank the optimal habitats for hares. However, the population crash of brown hare due to different epizootic diseases was also an important factor affecting the species' abundance and availability [46,47].

In contrast to the Pannonian population of the EIE [24], water birds and carrion remarkably reduced their ratio in the EIE's diet. After the country's accession to the EU (2006) and the introduction of strict sanitary regulations concerning carcass disposal [48], carrion became less frequented in the eagle's diet. The reduction of water birds was probably associated with a decrease in their abundance. It was recorded that some colonies of herons (Ardeidae) distributed along the lower reaches of the Tundzha river disappeared or reduced in number. On the other hand, wildfowl (Anatidae) that were more frequently predated in the winter period [22] decreased their abundance in the study area (author's data).

The reduction of gulls in the EIE's food spectrum was probably related to the reduction of their abundance, which could be a result of the elimination of unregulated landfills, concentrating large flocks of birds. However, yellow-legged gulls (*Larus michahelis*) are still one of the main food sources for eagles in the neighboring population of European Turkey [23].

The substitution of brown hare, water birds, and Poultry in some regions by northern white-breasted hedgehog and white stork could hardly be associated with a sharp increase in the abundance of these substitute prey species. The substitute prey probably existed in the territories of the eagles with similar abundance, but the eagles met their nutritional needs predating hares and easy catches, such as poultry species, which, being in significant quantities, represented a more nutritious source of biomass. Therefore, if these species decreased, eagles had to switch to another, less nutritious yet plentiful food source, such as hedgehogs, or more difficult to capture but with more biomass, such as white storks. However, this issue needs further clarification. Nevertheless, the described drastic and large-scale transformation in grassland habitats [41] and the direct mass extermination of hedgehogs by fast-moving shredders may soon lead to the depletion of this favorable food source. However, for how long hedgehogs will remain suitable prey for the EIE is highly questionable since habitat suitability is expected to become less favorable when habitat transformation affects large areas.

The shift from gulls to doves, the increased proportion of tortoises, and the intraguild predation were probably related to the eagles' adaptation to different food sources. However, individuals ranked these subsequent prey species differently, which was in line with the competitive refuge model according to the optimal foraging theory (OFT) [12,49,50]. In any case, these circumstances need further research.

4.2. *Profitable Prey Abundance Changes and Adaptive Response of EIE*

We found that the abundance of profitable prey for the EIE, such as sousliks, depleted trough the studied periods. This was clearly evident in the last period for most of the studied regions, except DHWstr, where souslik availability and abundance were very scarce and where its presence in the eagles' diet was less than 4% [22]. However, the decrease of souslik could be associated with the vast habitat alteration reported for the EIE distribution range in Bulgaria [41]. It is crucial to understand the particular diet response of eagles to the habitat changes in each occupied territory as well as whether this reaction depends on the size or any other characteristics of the favorable habitat. Anyway, this issue should be the focus of future research.

Our expectation that the presence of sousliks in the eagle's diet would follow the dynamics' pattern of this prey was confirmed. Despite the severe decline of the profitable prey, such as sousliks, the EIE population benefited in most of the studied regions, evidence for successful adaptation of this top predator. Our study confirmed previous findings, namely that the EIE was able to alter its diet and utilize the most available and/or abundant prey sources [22,24]. The significant shifting towards hedgehogs, white storks, pigeons, tortoises, and birds of prey was a good example of the successful adaptation of the EIE to a novel and accessible food source. Similar adaptation is known for another large top predator, the golden eagle (*Aquila chrysaetos*), which substitutes hedgehogs for its favorite prey, tortoises [16]. However, eagles could only shift and survive in those territories where their main prey decreased if alternative species were available and sufficiently abundant. For example, in parallel with the decrease of souslik populations, eagles' abundance also gradually declined in mountain regions (SG, ER). In fact, the last known mountain EIE's territory has been unoccupied since 2016. We speculate that the availability and abundance of substitute prey, such as hedgehogs, storks, and pigeons, was not enough to secure and sustain the birds, hence they abandoned these territories. Depression in the EIE population due to souslik degradation was reported in different regions of Russia [51,52]. However, this issue needs further confirmation.

Conversely, the decrease of sousliks, brown hare, and poultry in the rest of the study area forced eagles to prey more intensively on hedgehogs or forage for substitute prey, such as white stork, different reptiles, diurnal and nocturnal raptors, or songbirds. As a consequence, the EIE population expanded or remained stable in this part of the distribution area.

The trophic strategy used by eagles towards opportunistic foraging is an ecological advantage that allows the species to adapt to different habitats. According to the alternative prey hypothesis (APH), [53], a generalist predator such as the EIE may synchronize its diet with the fluctuations of main and alternative prey groups.

4.3. *Effects of Diet Alteration and Conservation Suggestions*

The importance of brown hare and the presence of small game species (phasants, partrige) in the prey of the EIE stirred a significant negative attitude among hunters towards eagles [24]. This effect intensified the "human–predator" conflict and led to human-related mortality due to persecution. Illegal shooting accounted for 12.5% of EIE mortality in Bulgaria [54], and there is evidence that this threat is increasing. It is crucial to communicate actively with and raise the conservation awareness of hunters. Improved communication between conservationists and hunters is known to be effective both in reducing violations and recognizing the mutual interest in lobbying for environmentally friendly practices in agricultural land use [24].

The increasing frequency of feral pigeon (*Columba livia var. domestica*) in the EIE's diet can also raise conflicts with pigeon fanciers, which in turn could result in persecution incidents. An immature eagle tagged with a satellite transmitter was poisoned through a bait set by pigeon fanciers in the second most important temporary settlement site of the species in the country [55]. In this area, poison baits set by pigeon fanciers cause mortality of different raptors such as long-legged buzzard (*Buteo rufinus*), peregrine falcon (*Falco peregrinus*), and saker falcon (*Falco cherrug*) [55,56].

The eagles' predation on poultry species, particularly intensive in the first study period, could also raise conflicts with poultry keepers, which would result in persecution incidents.

Feeding on carrion poses a potential threat of poisoning due to illegal baits used to control predators. Poisoning was identified as the most important mortality factor affecting the breeding population of the EIE in Bulgaria [54].

5. Conclusions

We found long-term and large-scale diet alterations of EIE. While brown hare, poultry, gulls, water birds, and carrion decreased over the years, northern white-breasted hedgehog and doves increased both in frequency and biomass provision. Raptors and owls raised their participation, but white stork and different reptiles supplied more biomass.

The abundance of European souslik decreased through the studied periods, which accounted for the lower proportion of this prey species in the eagle's diet. Nevertheless, the EIE population successfully adapted and significantly increased in most of the distribution area. Our idea that eagles could survive and expand in territories where their profitable prey decreased only if alternative species were available and abundant, was indirectly confirmed. The observed adaptive plasticity through alterations of the EIE's diet in response to temporal and spatial prey changes greatly facilitates conservation efforts, as it seems that although the species feeds on the most abundant prey, it does not depend solely on the state of any particular source of food. Therefore, conservation efforts should focus on the preservation of its main foraging habitats and the restoration of damaged ones so as to maintain a good condition of both its main food source in the area and the subsequent prey. Predator–prey interactions and conservationists–stakeholders conflict management are crucial for the effective preservation of this endangered top predator.

Supplementary Materials: The following supporting information can be downloaded at: https://www.mdpi.com/article/10.3390/d14111000/s1, Table S1: Results of the General Liner Mixed Model (GLMM) carried out to analyze the trend of the different prey categories (frequency and biomass) in different regions.

Author Contributions: Conceptualization, D.D. (Dimitar Demerdzhiev) and D.D. (Dobromir Dobrev); methodology, D.D. (Dimitar Demerdzhiev) and D.D. (Dobromir Dobrev); software, D.D. (Dimitar Demerdzhiev) and D.D. (Dobromir Dobrev); validation, D.D. (Dimitar Demerdzhiev), Z.B., D.D. (Dobromir Dobrev), N.N. and T.P.; formal analysis, D.D. (Dimitar Demerdzhiev); investigation, D.D. (Dimitar Demerdzhiev), Z.B., D.D. (Dobromir Dobrev), N.N. and T.P.; resources, D.D. (Dimitar Demerdzhiev); data curation, D.D. (Dimitar Demerdzhiev) Z.B., D.D. (Dobromir Dobrev), N.N. and T.P.; writing—original draft preparation, D.D. (Dimitar Demerdzhiev); writing—review and editing, Z.B., D.D. (Dobromir Dobrev), N.N. and T.P.; visualization, D.D. (Dimitar Demerdzhiev) and D.D. (Dobromir Dobrev); supervision, D.D. (Dimitar Demerdzhiev), D.D. (Dobromir Dobrev) and Z.B.; project administration, D.D. (Dobromir Dobrev); funding acquisition, D.D. (Dobromir Dobrev). All authors have read and agreed to the published version of the manuscript.

Funding: This work was partially funded by the LIFE Program of the European Union under the project 'Restoration and sustainable management of Imperial Eagle's foraging habitats in key Natura 2000 sites in Bulgaria' LIFE14 NAT/BG/001119.

Institutional Review Board Statement: Not applicable.

Data Availability Statement: Not applicable.

Acknowledgments: We would like to thank Georgi Popgeorgiev, Vladislav Vergilov, Nikolay Tzankov, and Andrey Stoyanov for their assistance with the identification of the remains of amphibians and reptiles; Vassil Popov and Nikolay Spasov for the identification of some mammalian remains; Mladen Jivkov and Tihomir Stefanov for the identification of fish prey. Our special thanks go to Nikolay Terziev for nest climbing and collection of a part of the food remains. We are grateful (alphabetically) to Atanas Delchev, Atanas Demerdzhiev, Aleksandar Georgiev, Dimitar Plachiyski, Georgi Georgiev, Georgi Gerdzhikov, Hristo Hristov, Iliya Iliev, Ivaylo Angelov, Kiril Metodiev, Krasimir Andonov, Krasimira Demerdzhieva, Milan Bakalov, Stefan Avramov, Stoycho Stoychev, Svetoslav Spasov, Vanyo Angelov, Vera Dyulgerska, Vladimir Dobrev, Vladimir Trifonov, and Volen Arkumarev, who took part in the field work. Without their assistance this survey would not have been possible. We are grateful to two anonymous reviewers who improved the draft of the manuscript.

Conflicts of Interest: The authors declare no conflict of interest.

References

1. Newton, I. *Population Limitation in Birds*; Academic Press: Cambridge, MA, USA, 1998; p. 597.
2. Reynolds, R.T.; Lambert, J.S.; Kay, S.L.; Sanderlin, J.S.; Bird, B.J. Factors affecting lifetime reproduction, long-term territory-specific reproduction, and estimation of habitat quality in northern goshawks. *PLoS ONE* **2019**, *14*, e0215841. [CrossRef] [PubMed]
3. Väli, Ü. Factors limiting reproductive performance and nestling sex ratio in the Lesser Spotted Eagle *Aquila pomarina* at the northern limit of its range: The impact of weather and prey abundance. *Acta Ornithol.* **2012**, *47*, 157–168. [CrossRef]
4. Lourenço, R.; Delgado, M.D.M.; Campioni, L.; Korpimäki, E.; Penteriani, V. Evaluating the influence of diet-related variables on breeding performance and home range behaviour of a top predator. *Popul. Ecol.* **2015**, *57*, 625–636. [CrossRef]
5. Senior, A.M.; Nakagawa, S.; Lihoreau, M.; Simpson, S.J.; Raubenheimer, D. An overlooked consequence of dietary mixing: A varied diet reduces interindividual variance of fitness. *Am. Nat.* **2015**, *186*, 649–659. [CrossRef]
6. Fatuyma, D.J.; Moreno, G. The evolution of ecological specialization. *Ann. Rev. Ecol. Syst.* **1988**, *19*, 207–233. [CrossRef]
7. Whitfield, D.P.; Reid, R.; Haworth, P.F.; Madders, M.; Marquiss, M.; Tingay, R.; Fielding, A.H. Diet specificity is not associated with increased reproductive performance of golden eagles *Aquila chrysaetos* in western Scotland. *Ibis* **2009**, *151*, 255–264. [CrossRef]
8. Bolnick, D.I.; Svanbäck, R.; Fordyce, J.A.; Yang, L.H.; Davis, J.M.; Hulsey, C.D.; Forister, M.L. The ecology of individuals: Incidence and implications of individual specialization. *Am. Nat.* **2003**, *161*, 1–28. [CrossRef]
9. Murgatroyd, M.; Avery, G.; Underhill, L.G.; Amar, A. Adaptability of a specialist predator: The effects of land use on diet diversification and breeding performance of Verreaux's eagles. *J. Avian Biol.* **2016**, *47*, 001–012. [CrossRef]
10. Moleón, M.; Sánchez-Zapata, J.A.; Gil-Sánchez, J.M.; Ballesteros-Duperón, E.; Barea-Azcón, J.M.; Virgós, E. Predator-prey relationships in a Mediterranean vertebrate system: Bonelli's eagles, rabbits and partridges. *Oecologia* **2012**, *168*, 679–689. [CrossRef]
11. Pires, M.M.; Guimarães, P.R., Jr.; Araújo, M.S.; Giaretta, A.A.; Costa, J.C.L.; dos Reis, S.F. The nested assembly of individual-resource networks. *J. Anim. Ecol.* **2011**, *80*, 896–903. [CrossRef]
12. Svanbäck, R.; Bolnick, D.I. Intraspecific competition affects the strength of individual specialization: An optimal diet theory method. *Evol. Ecol. Res.* **2005**, *7*, 993–1012.
13. Levin, S.A. The problem of pattern and scale in ecology. *Ecology* **1992**, *73*, 1943–1967. [CrossRef]
14. Hanski, I.; Henttonen, H.; Korpimäki, E.; Oksanen, L.; Turchin, P. Small-rodent dynamics and predation. *Ecology* **2001**, *82*, 1505–1520. [CrossRef]
15. Koenig, W.D.; Liebhold, A.M. Effects of periodical cicada emergences on abundance and synchrony of avian populations. *Ecology* **2005**, *86*, 1873–1882. [CrossRef]
16. Milchev, B. Diet shifting of tortoise-eating Golden Eagles (*Aquila chrysaetos*) in southeastern Bulgaria. *Ornis Fenn.* **2022**, *99*, 60–70. [CrossRef]
17. Moleón, M.; Sánchez-Zapata, J.A.; Real, J.; Garćia-Charton, J.A.; Gil-Sánchez, J.M.; Palma, L.; Bautista, J.; Bayle, P. Large-scale spatio-temporal shifts in the diet of a predator mediated by an emerging infectious disease of its main prey. *J. Biogeogr.* **2009**, *36*, 1502–1515. [CrossRef]
18. Angelstam, P.; Lindström, E.; Widen, P. Synchronous short-term population fluctuations of some birds and mammals of Fennoscandia—Occurrence and distribution. *Holarct. Ecol.* **1985**, *8*, 285–298. [CrossRef]
19. Moleón, M.; Almaraz, P.; Sánchez-Zapata, J.A. An emerging infectious disease triggering large-scale hyper predation. *PLoS ONE* **2008**, *3*, e2307. [CrossRef]
20. Del Hoyo, J.; Elliott, A.; Sargatal, J. *Handbook of the Birds of the World: New World Vultures to Guineafowl*; Lynx Edicions: Barcelona, Spain, 2007; Volume 2.
21. Chavko, J.; Danko, Š.; Obuch, J.; Mihók, J. The Food of the Imperial Eagle (*Aquila heliaca*) in Slovakia. *Slov. Raptor J.* **2007**, *1*, 1–18. [CrossRef]
22. Demerdzhiev, D.; Boev, Z.; Dobrev, D.; Terziev, N.; Nedyalkov, N.; Stoychev, S.; Petrov, T. Diet of Eastern Imperial Eagle (*Aquila heliaca*) in Bulgaria: Composition, distribution and variation. *Biodivers. Data J.* **2022**, *10*, e77746. [CrossRef]

23. Demerdzhiev, D.; Dobrev, D.; Isfendiyaroğlu, S.; Boev, Z.; Stoychev, S.; Terziev, N.; Spasov, S. Distribution, abundance, breeding parameters, threats, and prey preferences of the eastern imperial eagle (*Aquila heliaca*) in European Turkey. *Slov. Raptor J.* **2014**, *8*, 17–25. [CrossRef]
24. Horváth, M.; Solti, B.; Fatér, I.; Juhász, T.; Haraszthy, L.; Szitta, T.; Ballók, Z.; Pásztory-Kovács, S. Temporal changes in the diet composition of the Eastern Imperial Eagle (*Aquila heliaca*) in Hungary. *Ornis Hung.* **2018**, *1*, 1–26. [CrossRef]
25. Horváth, M.; Szitta, T.; Firmánszky, G.; Solti, B.; Kovács, A.; Moskát, C. Spatial variation in prey composition and its possible effect on reproductive success in an expanding Eastern Imperial Eagle (*Aquila heliaca*) population. *Acta Zool. Acad Sci. Hung.* **2010**, *56*, 187–200.
26. Katzner, T.; Bragin, E.; Knick, S.; Smith, A. Spatial structure in the diet of Eastern Imperial Eagle *Aquila heliaca* in Kazakhstan. *J. Avian Biol.* **2006**, *37*, 594–600. [CrossRef]
27. Karyakin, I.; Kovalenko, A.; Levin, A.; Pazhenkov, A. Eagles of the Aral–Caspian Region, Kazakhstan. *Raptors Conserv.* **2011**, *22*, 92–152.
28. Karyakin, I.; Nikolenko, E.; Levin, A.; Kovalenko, A. Imperial Eagle in Russia and Kazakhstan: Population status and trends. *Raptors Conserv.* **2008**, *14*, 18–27.
29. Abuladze, A. *Ecology of the Imperial Eagle* Aquila heliaca *in Georgia*; Meyburg, B.-U., Chancellor, R.D., Eds.; Eagle Studies; WWGBP: Berlin, Germany; London, UK; Paris, France, 1996; pp. 447–457.
30. Demerdzhiev, D. *Eastern Imperial Eagle (Aquila heliaca heliaca Savigny, 1809) (Accipitridae–Aves) in Bulgaria—Distribution, Biology, Ecology, Numbers, and Conservation Measures*; Synopsis of the Ph.D. Thesis; BAS-NMNH: Sofia, Bulgaria, 2011. (In Bulgarian, with English summary)
31. Katzner, T.; Bragin, E.; Knick, S.; Smith, A. Relationship between demographics and diet specificity of Eastern Imperial Eagle *Aquila heliaca* in Kazakhstan. *Ibis* **2005**, *147*, 576–586. [CrossRef]
32. Aulagner, S.; Haffner, R.; Mitchell-Jones, A.; Moutou, F.; Zima, J. *Mammals of Europe, North Africa and the Middle East*; Christopher Helm: London, UK, 2009.
33. Boev, Z. Comparative Morphological Studies of Herons (Family Ardeidae–Aves) from Bulgaria. Ph.D. Thesis, Sofia University St. Kliment Ohridski, Sofia, Bulgaria, 1986; 193p. (In Bulgarian).
34. Böhme, W. (Ed.) Askulapnatter (*E. longisima* Laurenti 1768). In *Handbuch der Reptilien und Amphibien Europas*; Aula Verlag: Weisbaden, Germany, 1993; pp. 331–372.
35. Dunning, J. (Ed.) *Handbook of Avian Body Masses*; CRC Press: Abingdon, UK, 2008.
36. Kunstmüller, I. The food ecology and the diet composition of Eagle Owl (*Bubo bubo*) at the breeding sites in the Bohemian-Moravian highlands, Czech Republic. *Crex* **2000**, *16*, 50–59.
37. Petrov, P. Variability in the body mass of European Hare (*Lepus europaeus*, Pall.) in Bulgaria. *Forest Sci.* **1964**, *1*, 73–81. (In Bulgarian)
38. Simeonov, S.; Michev, T.; Nankinov, D. *Fauna of Bulgaria: Aves, 1*; BAS: Sofia, Bulgaria, 1990; Volume 20. (In Bulgarian)
39. Simeonov, S.; Petrov, T.S. Studying the diet of Eastern Imperial Eagle (*Aquila heliaca* Savigny), Common Buzzard (*Buteo buteo* L.), and Rough-legged Buzzard (*Buteo lagopus* Brünnich) in Bulgaria. *Ecology* **1980**, *7*, 22–30. (In Bulgarian)
40. Koshev, Y. Distribution and status of the European Ground Squirrel (*Spermophilus citellus*) in Bulgaria. *Lynx* **2008**, *39*, 251–261.
41. Demerdzhiev, D.; Dobrev, D.; Popgeorgiev, G.; Stoychev, S. Landscape alteration affects the demography of an endangered avian predator by reducing the habitat quality. *Avian Res.* **2022**, *13*, 100030. [CrossRef]
42. Shapiro, S.; Wilk, M.; Chen, H. A comparative study of various tests of normality. *J. Am. Stat. Assoc.* **1968**, *63*, 1343–1372. [CrossRef]
43. StatSoft Inc. STATISTICA (Data Analysis Software System), Version 12. 2013. Available online: www.statsoft.com (accessed on 12 March 2019).
44. Marin, S.; Ivanov, I.; Georgiev, D.; Boev, Z. *On the Food of the Imperial Eagle Aquila heliaca on Sakar Mountain and Dervent Heights, Bulgaria*; Chancellor, R., Meyburg, B.-U., Eds.; Raptors Worldwide; WWGBP/MME: Towcester, UK, 2004; pp. 589–592.
45. Zhelev, C.H.; Ninov, N.; Mihaylov, H.; Gruychev, G.; Stoyanov, S.; Mirchev, R. Density of Brown Hare (*Lepus europaeus* Pallas, 1778) in the plain habitats of Bulgaria. In Proceedings of the 2nd International Symposium on Hunting, Modern Aspects of Sustainable Management of Game Populations, Novi Sad, Serbia, 17–20 October 2013.
46. Godfroid, J.; Cloeckaert, A.; Liautard, J.-P.; Kohler, S.; Fretin, D.; Walravens, K.; Garin-Bastuji, B.; Letesson, J.-J. From the discovery of the Malta fever's agent to the discovery of a marine mammal reservoir, brucellosis has continuously been a re-emerging zoonosis. *Vet. Res.* **2005**, *36*, 313–326. [CrossRef]
47. Wibbelt, G.; Frölich, K. Infectious diseases in European Brown Hare (*Lepus europaeus*). *Wildl. Biol. Pract.* **2005**, *1*, 86–93. [CrossRef]
48. Margalida, A.; Moleón, M. Toward carrion-free ecosystems? *Front. Ecol. Environ.* **2016**, *14*, 183–184.
49. Pyke, G.H.; Pulliam, R.; Charnov, E.L. Optimal foraging: A selective review of theory and tests. *Q. Rev. Biol.* **1977**, *52*, 137–154. [CrossRef]
50. Stephens, D.W.; Krebs, J.R. *Foraging Theory*; Princeton University Press: Princeton, NJ, USA, 1986.
51. Bekmansurov, R.H.; Karyakin, I.V. Results of Monitoring of the Imperial Eagle Population in the Republic of Tatarstan in 2011–2012, Russia. *Raptors Conserv. J.* **2013**, *26*, 84–108.
52. Sokolov, A.Y.; Sarychev, V.S.; Vlasov, A.A. Species of *Aquila* and *Haliaeetus* in the Breeding Fauna of the Central Black Earth Region: Current Status and Prospects for Existence. *Raptors Conserv. J.* **2019**, *38*, 109–126. [CrossRef]

53. Angelstam, P.; Lindström, E.; Widen, P. Role of predation in short-term population fluctuations of some birds and mammals in Fennoscandia. *Oecologia* **1984**, *62*, 199–208. [CrossRef]
54. Lazarova, I.; Dobrev, D.; Gradev, G.; Petrov, R.; Stoychev, S.; Klisurov, I.; Demerdzhiev, D. Main mortality factors for the Eastern Imperial Eagle (*Aquila heliaca* Savigny, 1809) in Bulgaria. *Ornis Hung.* **2020**, *28*, 120–134. [CrossRef]
55. Demerdzhiev, D. Birds in Besaparski Ridove special protection area (Natura 2000), Southern Bulgaria: Conservation status and dynamics. *Acta Zool. Bulg.* **2014**, *66* (Suppl. 5), 171–189.
56. Demerdzhiev, D. Breeding parameters and factors influencing the reproduction of an expanding Long-legged Buzzard (*Buteo rufinus*) population under high breeding density conditions. *J. Ornithol.* **2022**, *163*, 405–415. [CrossRef]

diversity

MDPI

Article

A Community-Level Approach to Species Conservation: A Case Study of *Falco naumanni* in Southern Italy

Alessandro Ferrarini *, Giuseppe Giglio, Stefania Caterina Pellegrino and Marco Gustin

Lipu-BirdLife Italy, Via Udine 3/a, I-43122 Parma, Italy; pinogiglio72@gmail.com (G.G.); stefaniacaterinapellegrino@gmail.com (S.C.P.); marco.gustin@lipu.it (M.G.)
* Correspondence: sgtpm@libero.it; Tel.: +39-0521-191-0728

Abstract: The conservation of threatened species is prevalently oriented towards two management strategies, i.e., habitat-level and species-level approaches. The former is focused on improving the conditions of the habitat of a certain species, whereas the latter is aimed at directly strengthening the species of interest. In this work, we adopted a different solution based on a community-level approach. Firstly, we identified the species (predators, competitors, prey) that interact with the species of interest (the lesser kestrel, *Falco naumanni*) in Southern Italy and mapped all of the ecological interactions among these species. Secondly, we built a simulation framework of the entire ecological network of the lesser kestrel. Thirdly, we simulated different management strategies that could increase the lesser kestrel population stock by targeting the species that interact with it. We found that the lesser kestrel in Southern Italy can be effectively protected by targeting the species interacting with the lesser kestrel, and that natural changes in the abundance of the interacting species could be used to pro-actively predict the dynamics of the lesser kestrel population. Our study demonstrates that a community-level approach to species conservation is highly appropriate on a local scale. Our methodological framework, based on qualitative modeling and what-if scenarios, can be applied in the absence of quantitative estimations of population stocks and interaction strengths.

Keywords: Alta Murgia; biotic community; decision framework; ecological network; lesser kestrel; management simulations; qualitative modeling; species interactions

Citation: Ferrarini, A.; Giglio, G.; Pellegrino, S.C.; Gustin, M. A Community-Level Approach to Species Conservation: A Case Study of *Falco naumanni* in Southern Italy. *Diversity* **2022**, *14*, 566. https://doi.org/10.3390/d14070566

Academic Editors: Vladimir D. Dobrev, Dobromir D. Dobrev and Dimitar A. Demerdzhiev

Received: 30 June 2022
Accepted: 12 July 2022
Published: 14 July 2022

1. Introduction

The lesser kestrel (*Falco naumanni*) is a colonial small falcon that breeds in the holes and fissures of urban buildings or farmhouses in the countryside [1]. This species prefers steppe-like grasslands and cultivated landscapes with short vegetation and extensive crops [2]. It is a spring and summer visitor in western Europe, migrating to Africa in autumn [3]. It is present in Annex I of the EU Wild Birds Directive (2009/147/EEC) and its important breeding habitats have been designated as SPAs (Special Protection Areas) of the Natura 2000 network in several nations, including Greece, Italy, Portugal, and Spain [4]. This species is also considered an important biological indicator for monitoring environmental changes because it is associated with high value habitats of the western Palaearctic [5], where it dwells in open ecosystems that are altered by agriculture.

Although it was once considered one of the most abundant raptors of the Palaearctic region, lesser kestrel populations have undergone substantial decline across Europe, its Asian range, and its African wintering grounds [3]. This population decline was primarily due to a reduction in the availability of nest sites, increases in the use of pesticides, and changes in land use [2,6]. In addition, many Palaearctic lesser kestrels are trans-Saharan migrants [7], and there is increasing evidence of the effects of climate change on their habitat associations, timing of arrival, and travel risks. Thus, the effects of climate change may add to those associated with local scale factors [8].

In Italy, the lesser kestrel has been widely studied in the Alta Murgia (Apulia region, Southern Italy) (Figure 1), an area that hosts the highest number of individuals in Italy in several urban colonies, and probably presents the highest densities of lesser kestrels in urban areas worldwide [9,10]. Recent studies have revealed the space use and flight attributes of this lesser kestrel population [11], its unexpected attitude towards frequent and long-distance nocturnal flights [12,13], and its elevated within-colony home range overlap and between-colony home range segregation [14–16].

Figure 1. Alta Murgia (Apulia and Basilicata regions, Southern Italy). The municipalities involved in this study are shown.

In this work, we first investigated the biotic community of the lesser kestrel in the Alta Murgia. We aimed to understand how changes (caused by external agents or by endogenous modifications) to the species interacting with the lesser kestrel can induce changes (both positive and negative) to its population stock. Hence, we used a qualitative modelling approach, which provided a method that is particularly useful whenever species and their interactions are known but not quantified [17,18]. Through simulations, we sought to detect (a) which interventions with the interacting species could help to preserve the lesser kestrel in the study area, and (b) how natural changes in the population parameters of the interacting species could be used to pro-actively predict the consequent impacts on the lesser kestrel population stock.

2. Materials and Methods

Firstly, based on our 10 years of field experience in the study area [9,10], we detected all of the species and/or taxonomic groups (players henceforth) that directly interact with the lesser kestrel. We detected six predators (the beech marten, the brown rat, the domestic cat, the magpie, the lanner falcon, and the peregrine falcon), three competitors (the European roller, the jackdaw, and the starling) and two kinds of prey (primary and secondary). Invertebrates (in particular, grasshoppers) are the primary prey of the lesser kestrel. In the absence or shortage of invertebrates, lesser kestrels feed on swifts, chiroptera, small reptiles, and small mammals (secondary prey).

Secondly, we split these players into two groups based on their frequent or infrequent interactions with lesser kestrels (Figure 2). The brown rat, the domestic cat, the magpie, the peregrine falcon, the jackdaw, and the invertebrates belonged to the former group (frequent players), whereas the beech marten, the lanner falcon, the European roller, the starling, and the secondary prey belonged to the latter group (infrequent players).

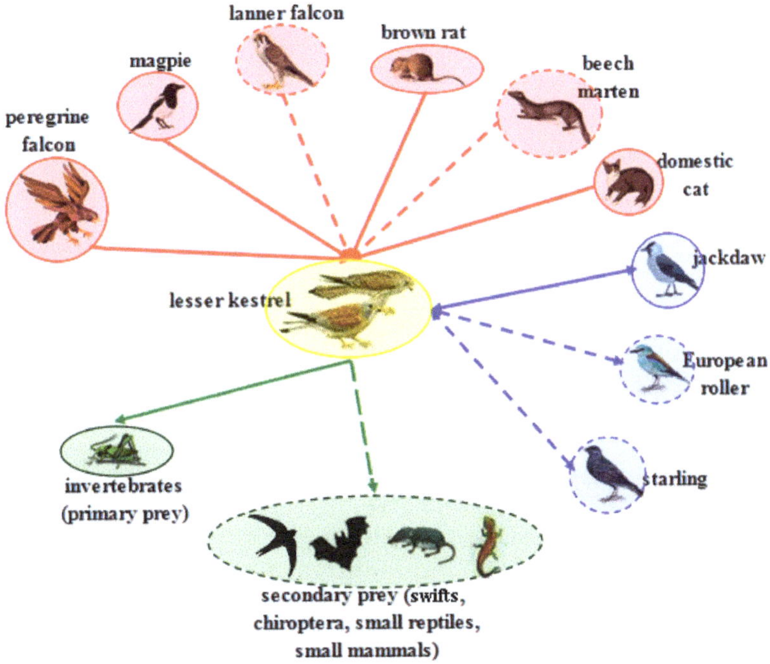

Figure 2. The biotic community of the lesser kestrel in the study area. The red, blue, and green arrows indicate the predators, the competitors, and the prey, respectively. The solid lines indicate frequent interactions and the dashed lines indicate infrequent interactions.

Thirdly, we built the current ecological network of the lesser kestrel in the study area, which represents all of the ecological interactions that are currently active among the frequent players and the species of interest.

Next, we built two what-if scenarios where we simulated the introduction of the two infrequent players (the beech marten and the secondary prey) into the current ecological network. We used these scenarios to determine whether a possible increase in the interactions between these infrequent players and the lesser kestrel could change the dynamics of the current ecological network. The latter scenario (secondary prey) simulated the future absence or shortage of invertebrates in the study area due to unfit agricultural practices (i.e., more intensive farming and an increased use of insecticides), which would force the lesser kestrels to shift their diet toward swifts, chiroptera, small reptiles, and small mammals (secondary prey). We avoided simulating alternative scenarios with the lanner falcon, the European roller, and the starling as, based on our 10 years of field experience, these species are rare in the study area, in fact the lanner falcon population is in decline. Therefore, the probability that these species become frequent players in the future is very low.

Next, we simulated the dynamics of the current scenario using qualitative modeling. The central feature of our methodological approach was that all ecological communities are composed of species connected to each other both directly and indirectly through a shared network of species interactions [19]. The presence of indirect effects implies that species need not interact directly in order to affect each other's populations, and that direct effects may be counteracted by indirect effects [20,21]. We used the loop analysis [22] as a means of encapsulating the topology of all the pairwise interactions present in the ecological network of the lesser kestrel in the Alta Murgia. The loop analysis is a qualitative modeling method that allows the evolution of the equilibrium values of system variables (i.e., species in our case study) to be predicted, following perturbations (inputs) that occur by permanently changing one or more parameters (e.g., mortality or fecundity) in the growth rate of the variables. Changes in the parameter values define new equilibrium points with new values for the levels of the variables. Inputs can be positive or negative and be caused by external agents or endogenous modifications. Because of the connections between components of the whole system, inputs propagate beyond their direct targets to all of the variables under study, and the reticulate interaction chains can magnify or dampen the effects of each direct interaction [22]. For example, indirect effects can coalesce and result in positive species responses even in systems where each species' direct effect on another is negative [23]. Any system variable can be the target of perturbations, and when an input (positive or negative) acts on a variable, there will be consequences both on the target variable and on the variables that directly or indirectly interact with it. In mathematical terms, each species' dynamics can be thought of as:

$$\frac{dX_j}{dt} = f_j(X_1, X_2, \ldots, X_n; c_1, c_2, \ldots, c_n) \quad (1)$$

where X_j is the variable for which the equilibrium value is calculated, the function f_j represents any imaginable relationship between the species X_j and the other species $X_{h \neq j}$ with population parameters $c_{h \neq j}$. In a loop analysis, the community matrix is a table where the rows represent the variables starting the effect and the columns represent the variables that receive the effect. Each h_j element of the community matrix represents the partial derivative of X_j, with respect to the growth rate of species h. The matrix values can be -1, 0, or 1. The loop analysis allows us to understand whether the equilibrium value of X_j (after simulating perturbations in the population parameters c_h of species $X_{h \neq j}$) is expected to increase (i.e., $dX_j/dt > 0$), decrease (i.e., $dX_j/dt < 0$), or remain the same (i.e., $dX_j/dt = 0$). In mathematical terms,

$$\frac{\partial X_j}{\partial c_h} = \frac{\sum_{h,k} \frac{\partial f_h}{\partial c_h} \times P_{jh}{}^{(k)} \times F_{n-k}{}^{(comp)}}{F_n} \quad (2)$$

where X_j is the variable (species) for which the equilibrium value is calculated; c_h is the parameter (growth rate) of the h_{th} species that is changing; $\partial f_h/\partial c_h$ describes whether the growth rate of the h_{th} variable increased or decreased after changing the parameter c_h; $P_{jh}{}^{(k)}$ is the product of the signs of the links comprising a causal pathway between variable X_j and X_h, including k variables; and $F_{n\text{-}k}{}^{(\text{comp})}$ represents feedback in the complementary subsystem of $P_{jh}{}^{(k)}$. F_n describes the overall feedback of the system, which is defined as:

$$F_n = \sum_{m=1}^{n} (-1)^{m+1} \times L(m, n) \quad (3)$$

where $L(m,n)$ is the product of the signs of m disjunct loops in n variables for $m = 1$ up to n, and $(-1)^{m+1}$ ensures that the sign of the feedback is corrected for n being even.

Equations (2) and (3) produce the table of predictions, i.e., a matrix whose entries denote the net effect (the sum of the direct and indirect effects) on variable j, resulting from a perturbation on variable h, i.e., the variations expected in the lesser kestrel population stock in response to positive/negative parameter inputs affecting any other species. Variations can be positive (population growth), negative (population decline), or null (when positive and negative effects tend to counteract each other, resulting in a net change of zero or a small change that is considered negligible). As the system variables are often connected to each other by multiple pathways, if such pathways have opposite effects, then the qualitative model can yield ambiguous predictions. Therefore, to safeguard against ambiguous predictions, we randomly assigned numerical values from a uniform distribution to the coefficients of the community matrix: -1 values were replaced with random values in the [−1, 0) interval and +1 values were replaced with random values in the (0, 1] interval. This was performed 100 × n^2 times, where n was the number of species. The frequency of positive or negative effects after these randomizations defined the sign and the probability of the net effect on variable j (i.e., the lesser kestrel), following a perturbation on variable h (e.g., the peregrine falcon). In order to meet the stability criterion, we only retained the simulations where the overall feedback (F_n) was negative [24].

We then repeated these simulations for the what-if scenarios.

3. Results

3.1. The Current Ecological Network of the Lesser Kestrel in the Alta Murgia

The current community matrix was a 7 × 7 table of the biotic interactions between the species that frequently interact with the lesser kestrel in the study area (Figure 3). The community matrix was dense. In fact, only 33.3% (i.e., 14 out of the 6 × 7 off-diagonal entries) of the pairwise interactions were equal to "no interaction" (i.e., blank cells in Figure 3). The predator–prey relationships were dominant (61.9%, i.e., 26 out of the 6 × 7 off-diagonal entries), whereas competition concerned only the lesser kestrel and the jackdaw.

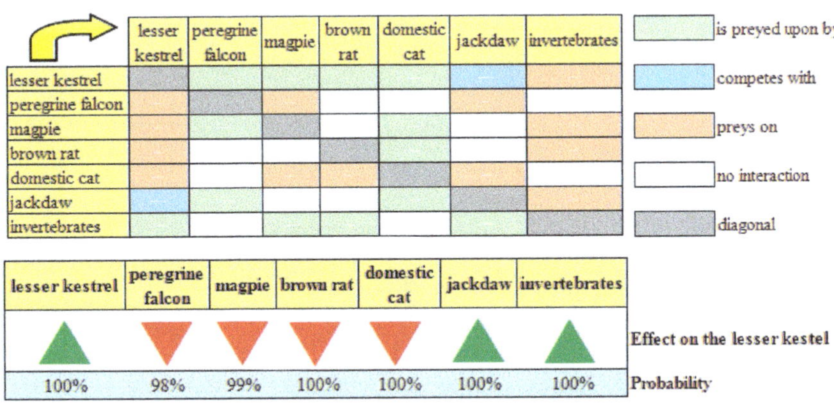

Figure 3. (**Top**) The current community matrix of the lesser kestrel in the Alta Murgia. The matrix must be read from the rows to the columns. (**Bottom**) Table of predictions. Key to reading: an increase in the growth rate of the peregrine falcon (for example, due to an increase in fecundity or a decrease in mortality) will result in a decrease in the lesser kestrel population stock, with probability equal to 98%. Predictions are shown for positive inputs. In the case of negative inputs (e.g., a decrease in the growth rate of the peregrine falcon), the predicted effects on the lesser kestrel population stock must be inverted, but the probabilities remain unchanged.

The table of predictions (Figure 3) showed unambiguous results. In fact, the net effects of changes in the growth rate of the frequent players on the lesser kestrel showed probabilities close or equal to 100% (out of 100 × 7^2 simulations, of which 812 were retained as the overall feedback (F_n) was <0, and thus the results were stable).

The lesser kestrel population stock was predicted to (a) increase with increased growth rates of the lesser kestrel, the jackdaw, and the invertebrates, and (b) decrease with decreased growth rates of these players. In contrast, the lesser kestrel population stock was expected to decrease with increased growth rates of the peregrine falcon, the magpie, the brown rat, and the domestic cat, and vice versa (i.e., to increase if the growth rates of these players decreased).

3.2. The "Beech Marten" Scenario

If the beech marten should become a frequent player, the ecological network of the lesser kestrel would change, as shown in Figure 4. The community matrix was dense. In fact, only 32.14% (i.e., 18 out of the 7 × 8 off-diagonal entries) of the pairwise interactions were equal to "no interaction" (i.e., blank cells in Figure 4). The predator–prey relations were dominant (64.28%, i.e., 36 out of the 7 × 8 off-diagonal entries), whereas competition concerned only the lesser kestrel and the jackdaw.

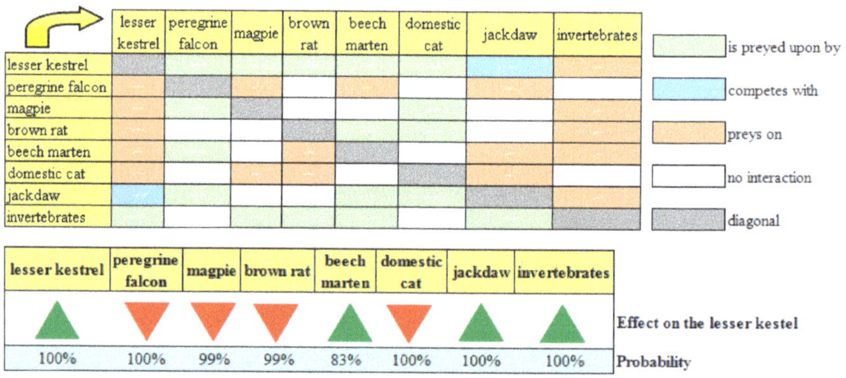

Figure 4. (**Top**) The community matrix of the lesser kestrel in the Alta Murgia for the "beech marten" scenario. The matrix must be read from the rows to the columns. (**Bottom**) Table of predictions. Key to reading: an increase in the growth rate of the jackdaw will result in an increase in the lesser kestrel population stock, with 100% probability. Predictions are shown for positive inputs. In the case of negative inputs (e.g., a decrease in the growth rate of the jackdaw), the predicted effects on the lesser kestrel population stock must be inverted, but the probabilities remain unchanged.

The table of predictions (Figure 4) showed that the presence of the beech marten would have an overall beneficial effect on the lesser kestrel. In fact, an increase in the growth rate of this species would have an 83% probability of determining an increase in the lesser kestrel population stock (out of 100 × 8^2 simulations, of which 860 were retained as the overall feedback (F_n) was <0, and thus the results were stable). With respect to the current scenario, the effects of the other species on the lesser kestrel remained unchanged.

3.3. The "Secondary Prey" Scenario

If the secondary prey replaced invertebrates in the lesser kestrels' diet, the ecological network of the lesser kestrel would be as shown in Figure 5. The community matrix was dense. In fact, only 23.81% (i.e., 10 out of the 6 × 7 off-diagonal entries) of the pairwise interactions were equal to "no interaction" (i.e., blank cells in Figure 5). The predator–prey relations were dominant (71.43%, i.e., 30 out of the 6 × 7 off-diagonal entries), whereas competition concerned only the lesser kestrel and the jackdaw.

Using 100 × 7^2 simulations (of which 930 were retained as the overall feedback (F_n) was <0, and thus the results were stable), the table of predictions (Figure 5) showed dynamics similar to those of the current ecological network. A notable exception was the peregrine falcon. In this scenario, an increase in the growth rate of this species would also increase the lesser kestrel population stock, and vice versa.

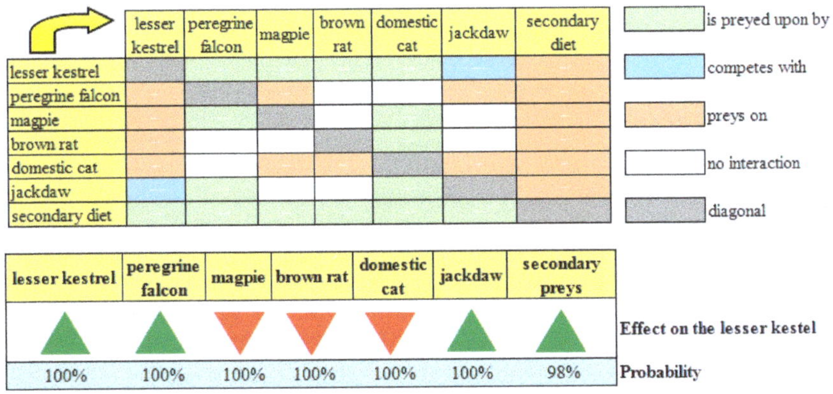

Figure 5. (**Top**) The community matrix of the lesser kestrel in the Alta Murgia for the "secondary prey" scenario. The matrix must be read from the rows to the columns. (**Bottom**) Table of predictions. Key to reading: an increase in the growth rate of the brown rat will result in a decrease in the lesser kestrel population stock, with probability equal to 100%. In the case of negative inputs (e.g., a decrease in the growth rate of the brown rat), the predicted effects on the lesser kestrel population stock must be inverted, but the probabilities remain unchanged.

4. Discussion

In this work, we have investigated the biotic community of the lesser kestrel in Southern Italy for the first time. We have used its ecological network to detect fit management options aimed at preserving this important bird species in a region that presents the highest density of lesser kestrels in urban areas worldwide. Our approach sought to distill the complexity of this biotic community down to manageable and measurable levels that allow reliable predictions of its behaviors [25,26]. The structure of the community interaction network was fundamental to our understanding of the factors that affect the lesser kestrel's dynamics through the propagation of different types of perturbations [27,28].

In the study area, the biotic community of the lesser kestrel is composed of nine species and two functional groups, i.e., primary and secondary prey. The peregrine falcon and the domestic cat are the top predators. The former preys on adult lesser kestrels and the latter preys on chicks. The beech marten, the brown rat, the domestic cat, the magpie, and the lanner falcon are intermediate predators. The beech marten and the brown rat feed on eggs, chicks, and adult lesser kestrels. The magpie feeds on eggs and chicks, whereas the lanner falcon preys on adult lesser kestrels. Two out of three biotic competitors (the jackdaw and the starling) compete with the lesser kestrels for the artificial nests that are placed on top of many buildings in the study area. The third competitor (the European roller) contends for food. In fact, it forages in the same areas (in particular, the pseudo-steppes, where invertebrates are more accessible and more abundant [29]) and sends the lesser kestrels away during their foraging attempts. The two types of prey can be considered alternative prey as the lesser kestrels largely prefer to feed on invertebrates. Thus, the secondary prey (swifts, chiroptera, small reptiles, and small mammals) enter the lesser kestrel's ecological network only when there is a substantial decline in invertebrate populations (in particular, grasshoppers) due to unfit agricultural practices (i.e., intensive agriculture with a large use of insecticides).

Currently, these interacting players present very different population stocks and trends. Based on our 10 years of field experience [9,10], the peregrine falcon and the brown rat are frequent and increasing in the study area, whereas the magpie, the jackdaw, the beech marten, and the domestic cat are frequent and stable. Although the beech marten is common in the study area, our field observations indicate that its interactions with the lesser kestrels are still infrequent. The lanner falcon is rare and its population is in decline, whereas the European roller and the starling are rare but their populations are slightly increasing. This qualitative assessment of the population stocks and trends required us to distinguish between the current biotic community (composed of the frequent players) and two alternative scenarios that could possibly occur in the future.

Interestingly, in all the simulated scenarios, two dynamics were shown to favor the lesser kestrel population stock with 100% probability. Any action aimed at increasing the lesser kestrel's growth rate will result in an increase in its population stock. Although this may seem obvious, it actually was not. In fact, a positive input on the lesser kestrel also favors its predators (the peregrine falcon, the magpie, the brown rat, the domestic cat, and the beech marten), which could increase their predation rates upon the lesser kestrel. However, due to the structural properties of the biotic community of the lesser kestrel, the net effect of an increase in the lesser kestrel's growth rate is expected to always be positive. Thus, any species-level conservation activity (e.g., the placement of new artificial nests atop urban buildings and farmhouses in order to increase the lesser kestrel's reproductive rates) will help to preserve the lesser kestrel population stock in the study area. The second type of action that favors the lesser kestrel population stock with 100% probability in all the simulated scenarios is purposed to increase the growth rate (and thus the abundance) of the primary prey. This may also seem obvious, however an increase in the growth rate of invertebrates will also favor the two predators (the magpie and the brown rat) and a competitor (the jackdaw) of the lesser kestrel. These species also feed on invertebrates (Figure 3). Due to the structural properties of the biotic community of the lesser kestrel, the net effect of these contrasting dynamics on the species of interest was positive in all of the simulated scenarios. Thus, a habitat-level approach based on (a) the preservation of the pseudo-steppes, (b) the optimal time for harvesting, and (c) the reduced use of pesticides will increase the abundance of invertebrates, which in turn will increase the lesser kestrel population stock. Of course, the opposite is also true. For example, if the artificial nests present in the study area are damaged or destroyed by meteorological events, or if agricultural practices become more intense, then the net effect on the lesser kestrel population stock will be negative with 100% probability.

The results of the simulations for the current scenario showed a counterintuitive result: an increase in the growth rate of the jackdaw would cause an increase in the lesser kestrel population stock (Figure 3). As the jackdaw is a competitor of the lesser kestrel, the opposite result was expected. However, an increase in the growth rate of the jackdaw will favor the domestic cat (which preys on the lesser kestrel, the magpie, the brown rat, and the jackdaw) and the peregrine falcon (which preys on the lesser kestrel, the magpie, and the jackdaw), and the net effect of these contrasting dynamics will be positive for the lesser kestrel with 100% probability. The results of the current scenario also showed several intuitive results: any action purposed to decrease (a) the growth rates of the brown rat and the magpie, or (b) the predation rate of the domestic cat will favor the lesser kestrel population stock. For example, frequent rodent control in the study area is desirable (i.e., using ad hoc traps and avoiding the use of rat poisons that could otherwise enter the food web of the lesser kestrel). In addition, the artificial nests used by the lesser kestrels should be placed out of reach of domestic cats, so as to lower their predation rate on young lesser kestrels. In contrast, as the peregrine falcon is a protected species in Italy, no action can be taken to decrease its growth rate. Thus, this species can only be used as an indirect indicator of the lesser kestrel population stock as it is negatively correlated to it.

Although the beech marten feeds on the lesser kestrel, the "beech marten" scenario showed that an increase in the population stock and growth rate of this species should not be impeded (Figure 4). In fact, the beech marten also feeds on the brown rat and the jackdaw, and the net effect of these contrasting dynamics will be positive for the lesser kestrel with 83% probability. In this scenario, the effects of the other species on the lesser kestrel would remain unchanged with respect to the current ecological network, including the positive effects of the jackdaw.

The "second prey" scenario showed another counterintuitive result (Figure 5): If there was a substantial decline in invertebrate populations (in particular, grasshoppers), an increase in the growth rate of the peregrine falcon would favor the lesser kestrel population stock. This would occur because the magpie, the brown rat, and the jackdaw would replace the invertebrates as the lesser kestrel's prey. However, in contrast to the current network, the secondary prey would also be preyed upon by the peregrine falcon and the domestic cat. Therefore, an increase in the growth rate of the peregrine falcon would subtract the food resources of two predators and one competitor of the lesser kestrel, and the net effect of these conflicting dynamics would be positive for the lesser kestrel with 100% probability. The peregrine falcon may also be useful to the lesser kestrel in the "starling" scenario as it is used to feed on the starling, which would become a frequent competitor of the lesser kestrel for the artificial nests that are placed on top of many buildings in the study area. This scenario was not simulated in this study because it is considered very improbable.

The methodological framework used in this study should be considered as a local-scale approach which cannot be generalized to regional or national areas because the ecological network of a certain species can change considerably from one area to another. For example, the lesser kestrel is also present in Northern Italy (in the Emilia-Romagna region) but, based on our field experience, its biotic community is very different in this region: the lanner falcon and the peregrine falcon are absent, however other species (the red-footed falcon, the common kestrel, the hooded crow, and the common buzzard) commonly interact with the lesser kestrel. In southern Portugal, the biotic community of the lesser kestrel was found to also include the barn owl and the little owl [30].

Qualitative modelling relies on the interactions between system components without quantitative data requirements [31]. Rather than focusing on known direct effects, qualitative modelling follows the pathways through which influences cascade in a biotic community [32]. While developing and using our qualitative modelling approach, a fundamental step was to detect the variables considered relevant and the interactions that represented how such variables influenced each other. Gaining insight into the ecological network of the study species is among the most fruitful results of qualitative modeling [33,34]. Our approach required neither detailed exact equations nor expensive field measurements, but instead focused on the qualitative character of the interactions (positive, negative, or zero) and the structure of the network. Using this approach, which sacrificed precision for generality and realism [35,36], we circumvented many of the uncertainties (e.g., observation error and variable functional responses) that impede quantitative models [37]. Recent applications of qualitative modelling to ecological and environmental systems are found in [38–42].

Although qualitative modelling makes simplified assumptions due to the absence of quantitative data, it allowed us to create a relatively complex picture of the dynamics of the lesser kestrel biotic community. By focusing on the direction of responses, our approach provided useful guidance about different management options without the need to collect quantitative data [43]. Firstly, a species-level approach purposed to directly increase the growth rate of this species will produce the desired effect. Secondly, a habitat-level approach aimed at preserving and extending the foraging habitats (where invertebrates are more abundant and easier to capture) will also increase the lesser kestrel population stock. As before, this outcome was not so obvious due to the complex structure of the ecological network of the lesser kestrel in the Alta Murgia. Thirdly, the presence of the jackdaw in the study area does not threaten the lesser kestrel because the net effect of its direct (negative) and indirect (positive) interactions with the lesser kestrel is positive. As the jackdaw is protected in Italy, favoring the presence of this species in the study area would produce two beneficial effects at the same time. Fourthly, the lesser kestrel would benefit from frequent rodent control. Fifthly, an increase in the abundance of beech marten in the study area should not be impeded. In fact, the net effect of its direct (negative) and indirect (positive) interactions with the lesser kestrel is positive. Sixthly, the interaction strength between the domestic cat and the lesser kestrel should be reduced, for example, by placing artificial nests where the domestic cats cannot reach. Lastly, the peregrine falcon is a threat to the lesser kestrel in the current biotic community, however, counteracting any decline in the lesser kestrel population stock would be essential in case invertebrate populations decrease in the study area due to unfit agricultural practices. In this instance, targeted actions that favor the growth rate of the peregrine falcon would help to increase the lesser kestrel population stock, despite its shift towards a secondary prey type.

5. Conclusions

Our study revealed a set of policies and management interventions that may help to preserve the lesser kestrel (*Falco naumanni*) in the Alta Murgia.

Overall, we created a decision framework for conservation of the species on a local-scale based on a data-poor approach that can be readily applied in the absence of quantitative estimations of population stocks and interaction strengths.

Author Contributions: Conceptualization, A.F., M.G., G.G. and S.C.P.; methodology, A.F.; software, A.F.; validation, M.G., G.G. and S.C.P.; formal analysis, A.F.; investigation, A.F., M.G., G.G. and S.C.P.; resources, M.G.; data curation, M.G., G.G. and S.C.P.; writing—original draft preparation, A.F.; writing—review and editing, M.G., G.G. and S.C.P.; visualization, A.F.; supervision, M.G.; project administration, M.G.; funding acquisition, M.G. All authors have read and agreed to the published version of the manuscript.

Funding: This study was funded by LIPU-UK.

Institutional Review Board Statement: Not applicable.

Informed Consent Statement: Not applicable.

Data Availability Statement: The dataset used in this study is available from the first author on reasonable request.

Acknowledgments: We thank Dino Scaravelli (University of Bologna, Italy) for his suggestions prior to submission. All individuals included in this section have consented to the acknowledgment. We thank the editors and two anonymous reviewers for their useful suggestions that improved this manuscript.

Conflicts of Interest: The authors declare no conflict of interest.

References

1. BirdLife International. *Birds in Europe: Population Estimates, Trends, and Conservation Status*; BirdLife International: Cambridge, UK, 2004.
2. BirdLife International. *European Birds of Conservation Concern: Populations, Trends and National Responsibilities*; BirdLife International: Cambridge, UK, 2017.
3. BirdLife International. *Falco naumanni. The IUCN Red List of Threatened Species: E.T22696357A40742561*; BirdLife International: Cambridge, UK, 2013.
4. BirdLife International. *European Red List of Birds*; Office for Official Publications of the European Communities: Luxembourg, 2015.
5. Cramp, S.; Simmons, K.E.L. *The Birds of the Western Palaearctic*; Oxford University Press: NewYork, NY, USA, 1980; Volume 2.
6. Bustamante, J. Predictive models for lesser kestrel *Falco naumanni* distribution, abundance and extinction in southern Spain. *Biol. Conserv.* **1997**, *80*, 153–160. [CrossRef]
7. Rodríguez, A.; Negro, J.J.; Bustamante, J.; Fox, J.W.; Afanasyev, V. Geolocators map the wintering grounds of threatened lesser kestrels in Africa. *Divers. Distrib.* **2009**, *15*, 1010–1016. [CrossRef]
8. Jetz, W.; Wilcove, D.S.; Dobson, A.P. Projected impacts of climate and land-use change on the global diversity of birds. *PLoS Biol.* **2007**, *5*, 1211–1219. [CrossRef] [PubMed]
9. Gustin, M.; Ferrarini, A.; Giglio, G.; Pellegrino, S.C.; Scaravelli, D. *Il Parco per il Grillaio (Falco naumanni) nel Parco Nazionale dell'Alta Murgia*; Lipu Press: Parma, Italy, 2013.
10. Gustin, M.; Ferrarini, A.; Giglio, G.; Pellegrino, S.C.; Scaravelli, D. Il Grillaio nel Parco Nazionale dell'Alta Murgia. Lipu Press: Parma, Italy, 2017.
11. Gustin, M.; Giglio, G.; Pellegrino, S.C.; Frassanito, A.; Ferrarini, A. Space use and flight attributes of breeding Lesser Kestrels *Falco naumanni* revealed by GPS tracking. *Bird Study* **2017**, *64*, 274–277. [CrossRef]

12. Gustin, M.; Giglio, G.; Pellegrino, S.C.; Frassanito, A.; Ferrarini, A. First evidences of nocturnal widespread activity of Lesser Kestrel *Falco naumanni* in Southern Italy. *Ornis Fenn.* **2014**, *91*, 256–260.
13. Gustin, M.; Giglio, G.; Pellegrino, S.C.; Frassanito, A.; Ferrarini, A. New evidences confirm that during the breeding season Lesser Kestrel is not a strictly diurnal raptor. *Ornis Fenn.* **2017**, *94*, 194–199.
14. Ferrarini, A.; Giglio, G.; Pellegrino, S.C.; Frassanito, A.; Gustin, M. First evidence of mutually exclusive home ranges in the two main colonies of Lesser Kestrels in Italy. *Ardea* **2018**, *106*, 85–89. [CrossRef]
15. Ferrarini, A.; Giglio, G.; Pellegrino, S.C.; Gustin, M. A new general index of home range overlap and segregation: The Lesser Kestrel in Southern Italy as a case study. *Avian Res.* **2021**, *12*, 4. [CrossRef]
16. Ferrarini, A.; Giglio, G.; Pellegrino, S.C.; Gustin, M. Measuring the Degree of Overlap and Segregation among Multiple Probabilistic Home Ranges: A New Index with Illustrative Application to the Lesser Kestrel *Falco naumanni*. *Animals* **2021**, *11*, 2913. [CrossRef]
17. Dambacher, J.M.; Li, H.W.; Rossignol, P.A. Qualitative predictions in model ecosystems. *Ecol. Model.* **2003**, *161*, 79–93. [CrossRef]
18. Ortiz, M.; Rodriguez-Zaragosa, F.; Hermosillo-Nunez, B.; Jordán, F. Control strategy scenarios for the alien lionfsh *Pterois volitans* in Chinchorro Bank (Mexican Caribbean) based on semi-quantitative loop network analysis. *PLoS ONE* **2015**, *10*, e0130261. [CrossRef]
19. Raymond, B.; McInnes, J.; Dambacher, J.M.; Way, S.; Bergstrom, D.M. Qualitative modelling of invasive species eradication on subantarctic Macquarie Island. *J. Appl. Ecol.* **2011**, *48*, 181–191. [CrossRef]
20. Polis, G.A.; Strong, D.R. Food web complexity and community dynamics. *Am. Nat.* **1996**, *147*, 813–846. [CrossRef]
21. Polis, G.A.; Winemiller, K.O. *Food Webs: Integration of Patterns and Dynamics*; Chapman and Hall: New York, NY, USA, 1996.
22. Puccia, C.J.; Levins, R. *Qualitative Modeling of Complex Systems*; Harvard University Press: Cambridge, MA, USA, 1985.
23. Roberts, A.; Stone, L. Advantageous indirect interactions in systems of competition. *J. Theor. Biol.* **2004**, *228*, 367–375. [CrossRef]
24. Dambacher, J.M.; Luh, H.K.; Li, H.W.; Rossignol, P.A. Qualitative stability and ambiguity in model ecosystems. *Am. Nat.* **2003**, *161*, 876–888. [CrossRef]
25. May, R.M. Will a large complex system be stable? *Nature* **1972**, *238*, 413–414. [CrossRef]
26. Bascompte, J. Disentangling the web of life. *Science* **2009**, *325*, 416–419. [CrossRef]
27. Doak, D.F.; Estes, J.A.; Halpern, B.S.; Jacob, U.; Lindberg, D.R.; Lovvorn, J.; Monson, D.; Tinker, M.; Williams, T.M.; Wootton, T.; et al. Understanding and predicting ecological dynamics: Are major surprises inevitable? *Ecology* **2008**, *89*, 952–961. [CrossRef]
28. Stone, L.; Roberts, A. Conditions for a species to gain advantage from the presence of competitors. *Ecology* **1991**, *72*, 1964–1972. [CrossRef]
29. Gustin, M.; Giglio, G.; Pellegrino, S.C.; Frassanito, A.; Ferrarini, A. Detected foraging strategies and consequent conservation policies of the Lesser Kestrel Falco naumanni in Southern Italy. *Proc. Int. Acad. Ecol. Environ. Sci.* **2014**, *4*, 148–161.
30. Catry, T.; Catry, I. Nest-site provisioning re-shapes species interactions within bird assemblages. *Ibis* **2019**, *161*, 699–704. [CrossRef]
31. Novak, M.; Wootton, J.T.; Doak, D.F.; Emmerson, M.; Estes, J.A.; Tinker, M.T. Predicting community responses to perturbations in the face of imperfect knowledge and network complexity. *Ecology* **2011**, *92*, 836–846. [CrossRef]
32. Justus, J. Qualitative scientific modeling and loop analysis. *Philos. Sci.* **2005**, *72*, 1272–1286. [CrossRef]
33. Dambacher, J.M.; Li, H.W.; Rossignol, P.A. Relevance of community structure in assessing indeterminacy of ecological predictions. *Ecology* **2002**, *83*, 1372–1385. [CrossRef]
34. Nuttle, T.; Bredeweg, B.; Salles, P.; Neumann, M. Representing and managing uncertainty in qualitative ecological models. *Ecol. Inf.* **2009**, *4*, 358–366. [CrossRef]
35. Levins, R. The strategy of model building in population biology. *Am. Sci.* **1966**, *54*, 421–431.
36. Levins, R. *Evolution in Changing Environments*; Princeton Monograph Series: Princeton, NJ, USA, 1968.
37. Levins, R. Qualitative mathematics for understanding, prediction, and intervention in complex ecosystems. In *Ecosystem Health*; Rapport, D., Costanza, R., Epstein, P., Gaudet, C., Levins, R., Eds.; Blackwell Scientific Publishers: Boston, MA, USA, 1998; pp. 178–204.
38. Alexandridis, N.; Bacher, C.; Jean, F.; Dambacher, J.M. Revealing perturbation responses with limited observations of biological communities. *Ecol. Indic.* **2021**, *128*, 107840. [CrossRef]
39. Abbott, K.C.; Eppinga, M.B.; Umbanhowar, J.; Baudena, M.; Bever, J.D. Microbiome influence on host community dynamics: Conceptual integration of microbiome feedback with classical host–microbe theory. *Ecol. Lett.* **2021**, *24*, 2796–2811. [CrossRef]
40. Ortiz, M.; Uribe, R.A. Comparing the stability of successional transient models in kelp forests and barrens (south-east Pacific): Implications for conservation monitoring. *Aquat. Conserv. Mar. Freshw. Ecosyst.* **2021**, *31*, 1297–1310. [CrossRef]
41. Reum, J.C.P.; McDonald, P.S.; Long, W.C.; Holsman, K.K.; Divine, L.; Armstrong, D.; Armstrong, J. Rapid assessment of management options for promoting stock rebuilding in data-poor species under climate change. *Conserv. Biol.* **2019**, *34*, 611–621. [CrossRef]
42. Sobocinski, K.L.; Greene, C.M.; Schmidt, M.W. Using a qualitative model to explore the impacts of ecosystem and anthropogenic drivers upon declining marine survival in Pacific salmon. *Environ. Conserv.* **2018**, *45*, 278–290. [CrossRef]
43. Dambacher, J.M.; Rothlisberg, P.C.; Loneragan, N.R. Qualitative mathematical models to support ecosystem-based management of Australia's northern prawn fishery. *Ecol. Appl.* **2015**, *25*, 278–298. [CrossRef] [PubMed]

diversity MDPI

Article

Importance of Mesohabitat for Nest-Site Selection in Breeding Eagle Owls (*Bubo bubo*): A Multi-Scale Model

Ezra Hadad [1], Dan Malkinson [2,3], Reuven Yosef [4,*], Gilad Weil [1] and Motti Charter [2,3]

1 Israel Nature and Parks Authority, 3 Am Ve'Olamo St., Jerusalem 9546303, Israel; ezra.hadad9@gmail.com (E.H.); weilgilad@gmail.com (G.W.)
2 Department of Geography and Environmental Studies, University of Haifa, Haifa 3498838, Israel; dmalk@geo.haifa.ac.il (D.M.); mcharter@geo.haifa.ac.il (M.C.)
3 Shamir Research Institute, Katzrin 1290000, Israel
4 Eilat Campus, Ben Gurion University of the Negev Eilat Campus, Eilat 8810201, Israel
* Correspondence: ryosef60@gmail.com; Tel.: +972-537671290

Abstract: Apex predators make excellent bioindicators of habitat quality and anthropogenic changes. Eagle owls (*Bubo bubo*) are such apex and keystone predators, who show preferential nest-site selection, usually away from human activities and habitation. However, Israel is a small country with a burgeoning human population. Hence, in order to understand the habitat requirements of eagle owls in central Israel, we conducted a multi-scale model analysis on their existing nest sites between 2006 and 2010. We identified 203 successful breeding attempts at 73 different sites. Our data suggested that the breeding population of the eagle owls was limited by the availability of quality nest locations, i.e., quarries, and caves. The probability of an eagle pair breeding increased with the presence of both quarries and caves, but was not related to mesohabitat properties. In addition, eagle owl breeding densities were positively related to the number of nest localities and to the planted woodlands. Furthermore, we found that eagle owls successfully raised their young regardless of the surrounding mesohabitat and sought the presence of other potential nest sites in the vicinity of the active nest sites, most likely due to the owls' opportunistic and generalist hunting behavior, which facilitated the consumption of a wide prey base. Appropriate nest sites (quarries and caves) appeared to increase population numbers and, therefore, should be protected Further studies should determine whether increasing artificial nest sites and reclaiming abandoned quarries could increase eagle owl numbers in a sustainable manner.

Keywords: nest site; limiting factors; quarry; cave; mesohabitat; eagle owl

Citation: Hadad, E.; Malkinson, D.; Yosef, R.; Weil, G.; Charter, M. Importance of Mesohabitat for Nest-Site Selection in Breeding Eagle Owls (*Bubo bubo*): A Multi-Scale Model. *Diversity* **2022**, *14*, 438. https://doi.org/10.3390/d14060438

Academic Editors: Vladimir D. Dobrev, Dobromir D. Dobrev, Dimitar A. Demerdzhiev and Michael Wink

Received: 2 April 2022
Accepted: 25 May 2022
Published: 30 May 2022

1. Introduction

Apex predator populations are important for biodiversity conservation and for regulating the effect of competition by mesopredators on prey availability [1]. As human populations expand and habitats are destroyed, apex predator population numbers—which were small to begin with—and the predators' ecological function are at risk [2,3].

Different factors such as nest-site availability and diet limit the populations of birds of prey [4,5]. To protect keystone species such as eagle owls (*Bubo bubo*), it is important to determine what factors affect their nest-site selection and breeding success. Previous studies reported that the occupation of a nest site by eagle owls may be affected by the proportion of specific mesohabitat properties surrounding the nest site, such as open landscapes [6], forests [7], protected areas [8], the distance to the road [7,9], intra-specific competition [7,10] and the elevation of the locality [11]. Similarly, eagle owl breeding success (i.e., the number of young fledged/pair) was also affected by surrounding mesohabitat properties such as [6,12] latitude [13] and elevation [14].

Even though mesohabitat properties/categories are important for some eagle owl populations, it is still unclear whether a lack of availability of nest sites can limit population numbers [15,16]. Eagle owls do not build nests, but lay eggs in a wide variety of nest sites [17,18] and frequently breed on the ground in many mesohabitat types (forests, cliffs, caves and quarries; [19]), but also avoid areas where human activity is high [8]. We assumed that just as nest cavities limit second-cavity bird breeding [4], specific nest sites may limit eagle owl breeding success and may therefore influence the persistence of breeding populations even more than the mesohabitat properties around the nest site themselves.

Protecting and conserving cryptic and elusive apex predators such as eagle owls are sometimes not prioritized in many conservation schemes [20]. In the case of eagle owls, this may result from the fact that their direct and indirect predatory ecosystem effects are difficult to observe and measure, due to their nocturnal lifestyles. To conserve the owls, it is of high importance to determine what factors limit the number of pairs, breeding success and the subsequent persistence of the population in the wild.

We studied a breeding population of eagle owls in Israel that utilized two types of habitats—quarries and caves—as nest sites that were not distributed evenly throughout the study region and therefore potentially limited breeding numbers. Specifically, we studied whether the availability of nesting sites and presence of different mesohabitats affected breeding densities, the probability of breeding and breeding success in eagle owls. We hypothesized that both nest-site types, but not mesohabitat availability around the nest sites, would affect the number of breeding pairs, but not the breeding success, due the eagle owls' opportunistic and generalist hunting behavior.

2. Materials and Methods

2.1. Study Area

The study was conducted in a 2644 km^2 area of the Judea region of Israel (central–southern Israel, 31°44′44.47″ N, 34°59′11.93″ E) during the 2006–2010 breeding seasons, because of a preliminary, unpublished survey conducted in the region and because it is the work area of Ezra Hadad—the ranger for the Israel Nature & Parks Authority (Figure 1). The area is semi-arid, with the average annual rainfall during the 2006–2010 breeding season from 15 February to 15 July each year measuring 439 mm (N = 5 years, SE = 44.3 mm), with a mean daily maximum temperature of 27.2 °C (N = 3 years, SE = 0.5) and a mean daily minimum temperature of 12.5 °C (SE = 0.7; Israel Meteorological Center).

The lithology in the study area consists mainly of karstic carbonates as well as shales of the Cretaceous Judea group [21] and is characterized by artificial bell-cave structures, particularly in the shale formations. In a research survey, the caves were discovered as occurring in clusters and were formed under shallow phreatic conditions prior to the major uplift of the terrain of central Israel during the late Cenozoic. The artificial bell caves are the result of the quarrying of blocks of chalk, used for building during the late Roman, Byzantine and early-Islamic periods [22].

The study area was visited from March to August each year during the 2006–2010 breeding seasons, at a frequency of once to twice a week. Nests were found by searches on foot for related signs such as the presence of adults, fresh pellets, vocalizations, etc. We assumed that all abandoned historical nest sites were unused for that season and those with either eggs or nestlings were assumed to be occupied sites. Because eagle owls are evenly spaced over a landscape to avoid territorial conflict [19], we conducted optimal stratified surveys [23–25] based on field experience and many years of field work at the study site. To prevent any disturbances to the breeding pairs, initial observations were conducted from a distance, using binoculars (Swarovski 10 × 42, Absam, Austria). The nests were visited weekly on foot after the nestlings hatched in order to determine breeding success, i.e., the number of young fledged/pair. All active quarries, caves and the location of active nest sites were recorded in the field using the Israel TM Grid coordinate system and uploaded to ARCMAP 10.4 (ESRI, Redlands, CA, USA) layer for the 2006–2010 breeding seasons (Figure 1).

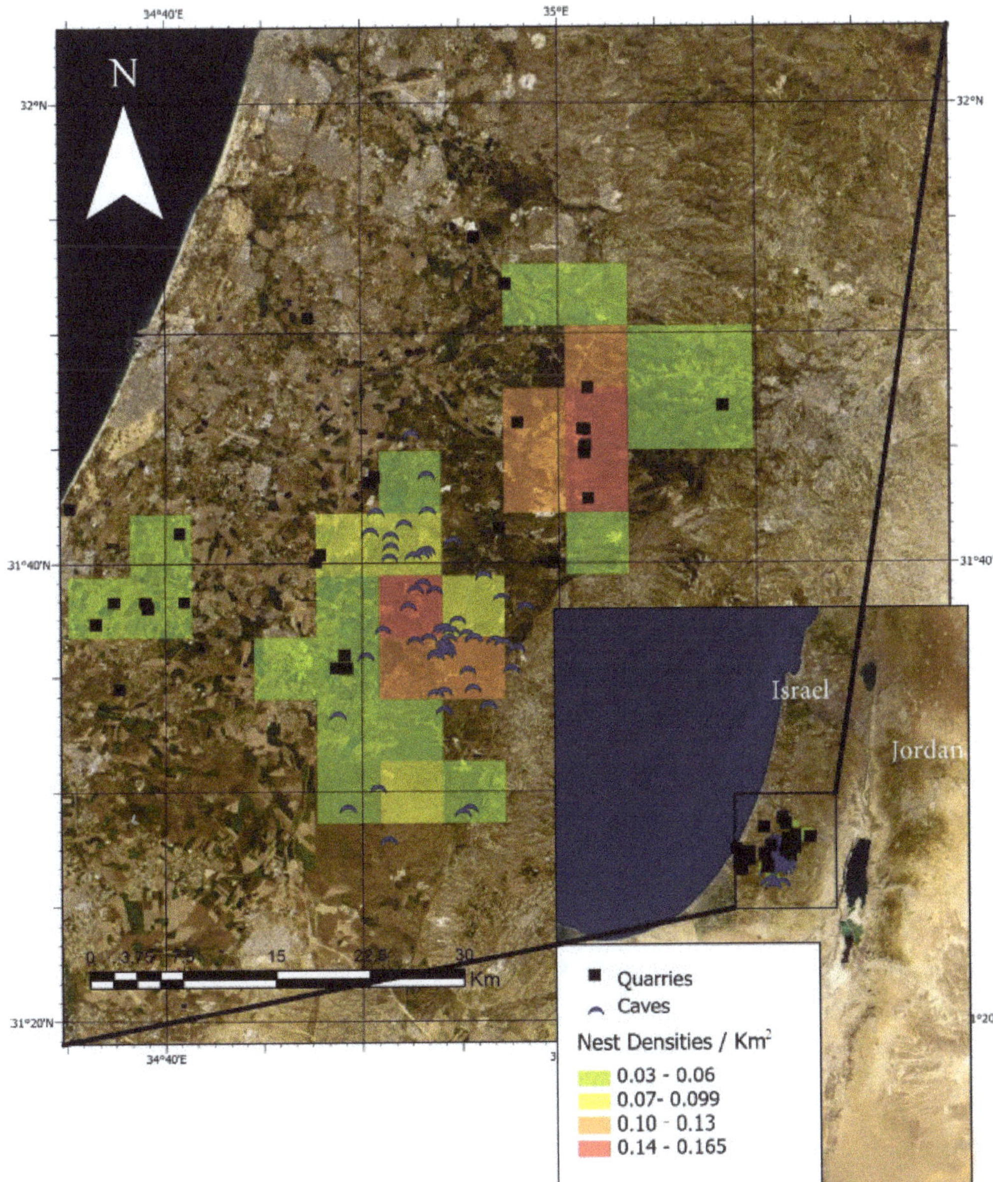

Figure 1. Map of the study site, the Judea region, in central Israel. The black squares denote the locations of all the quarries including those not occupied, the elliptical line denotes the bell caves and the colored blocks the relative densities of eagle owl (*Bubo bubo*) nest sites. The inset map shows the study area in relation to the region; each grid is 15 km^2. Not all the quarries and caves were occupied.

2.2. Statistical Analysis

For the 2006–2010 breeding data, we used a one-way ANOVA to compare the number of nestlings fledged throughout the years as well as a χ^2 test to compare the occupation of quarries and caves by the breeding pairs. We analyzed the number of potential nest localities (i.e., quarries or caves available but not occupied) around active nests and the proportion of different mesohabitat land cover categories around active nests (Table 1) to predict the use of nest localities, occupation of nest sites, and breeding success. We analyzed the 2006–2010 agglomerated breeding data both generally and for each year separately.

Table 1. Average proportion of land cover categories in spatial scales of 1000–5000 m radius around nest sites of eagle owls (*Bubo bubo*).

Scale (km)	1000	2000	3000	4000	5000
Built	8.4	9.1	10.4	11.4	12.1
Disturbed	2.3	1.5	1.1	1.1	1.1
Grasslands	0.8	1	1.1	1.2	1.3
Maquis	10.5	9.8	9.2	8.6	8.2
Shrub	12.4	10.9	10.2	9.9	9.6
Grove	22.4	21.7	22.1	22.2	21.9
Woods	19.4	18	16.6	15.9	15.5
Agriculture	18.3	21.5	22.5	23	23
Orchards	5.4	6.5	6.7	6.6	7.1
Water	0.1	0.1	0.1	0.2	0.2

The size of the home ranges is still uncertain. This may be because home ranges can vary yearly and most studies of eagle owl home ranges used VHF tracking [26–28], which typically underestimated the ranges, due to a low number of localizations as well as biased overestimations for both the locations taclose to the nest site and the locations of last sightings [29]. We therefore suggested increments of up to 5 km in distance (a 1–5 km radius) around active nests [30,31] that eagle owls could easily fly to during nocturnal foraging trips.

The mesohabitat land cover around the nest sites was defined as the proportion of each type of the land-use category (data from Israel's National Ecosystem Assessment Program, HaMaarag, Tel Aviv, Israel), using ARCMAP 10.4 (ESRI, Redlands, CA, USA). The mesohabitat land-use categories used for this analysis were anthropogenic-disturbed areas (including quarries and landfills), built areas (villages, industrial areas and buildings), water sources, grassland, agricultural field crops, planted woodlands, shrub lands, maquis and agricultural orchards (Table 1). All the caves and quarries were identified by geographical information systems (GIS) and ground-proofed in the field.

We applied logistic regression to assess the spatial distribution and location of the 73 nest sites found during the study, in relation to the quarries, caves and the mesohabitat land cover. We compared the spatial properties in relation to the location of the nest localities and the mesohabitat around the observed nest sites, to 73 randomly generated nest locations, utilizing the same buffer sizes (a 1–5 km radius) within the boundaries of the studied landscape.

Nest-site occupancy (binary response variable = nest sites occupied/not occupied) were compared during each of the study years, in relation to the number of nest localities and mesohabitat land cover. We used logistic regression to determine whether nest occupancy, during each year and at each of the scales, exhibited statistical dependence with respect to the above-mentioned explanatory variables.

We used a linear, mixed-model analysis [32] to determine whether nest density (the number of active nest sites in a 5 km radius, 78.54 km^2) was related to the number of potential nest localities (i.e., quarry and caves) and mesohabitat land uses. We first generated a random series of eight five km radius random buffers and repeated the procedure 10 times for each year, separately. We then calculated the number of potential nest localities, number of active nest sites and mesohabitat land cover within each buffer. We repeated the same buffer randomizations per year and added them as random factors, together with the year, to avoid overlap among the buffers and thus avoid pseudo-replication issues.

We performed statistical analyses using JMP 13.0, North Carolina, USA and SPSS version 22, Chicago, IL, USA.

3. Results

During the 2006–2010 breeding seasons, we located a total of 203 breeding attempts in 73 different nest sites (mean = 40.6 nest per year, SE = 0.3, N = 5, Figure 1) and found that the number of nestlings fledged per breeding pair (mean 2.81 = nestlings per year, SD = 0.83, N = 203 nests) did not significantly vary between the years ($F_{4,198} = 0.25$, $p = 0.91$; Figure 2). During 2006, we studied 35 EO breeding pairs; in 2007, we found 39 pairs; in 2008, we found 46 pairs; in 2009, there were 42 pairs and in 2010, we followed the breeding attempts of 41 pairs. The number of fledglings per breeding pair was: 2006—2.49 fledglings (SD = 0.92), 2007—2.44 (SD = 0.68), 2008—2.59 (SD = 0.83), 2009—2.57 (SD = 0.80) and 2010—2.59 (SD = 1.0) and found that the number of nestlings fledged (mean 2.81 = nestlings per year, SD = 0.83, N = 203 nests) did not significantly vary between the years ($F_{4,198} = 0.25$, $p = 0.91$; Figure 2).

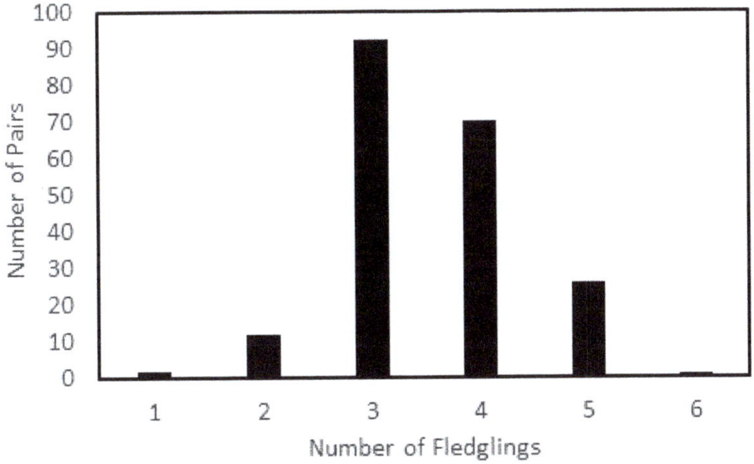

Figure 2. The number of fledglings of the eagle owl (*Bubo bubo*) during the years 2006–2010 in the Judea region of central Israel.

Throughout the study period, eagle owls preferentially occupied a higher percentage of quarries than caves (2006 – $\chi^2 = 11.5$, df = 1, $p < 0.001$; 2007 – $\chi^2 = 7.3$, df = 1, $p = 0.007$; 2008 – $\chi^2 = 15.0$, df = 1, $p < 0.001$; 2009 – $\chi^2 = 14.2$, df = 1, $p < 0.004$; 2010 – $\chi^2 = 9.0$, df = 1, $p = 0.003$; Figure 3).

The spatial distribution and location of the 73 nest sites found, compared to the 73 randomly selected locations, suggested that nest locations were significantly associated with quarries and caves (Table 2). The number of caves and/or quarries around each observed nest was higher compared to the randomly generated locations. For example, the mean number of observed caves at the 1 km buffer scale was 0.72, compared to the randomly simulated caves, which averaged 0.068. Similarly, at the five km buffer, the mean number of observed caves was 5.45, compared to a mean of 2.42 randomly simulated caves. Consistently, the association of the nest locations was stronger with quarries as compared to caves at all spatial scales. While being significant at all scales, the significance tended to decrease with the increase in scale. For example, the significance of the number of surrounding caves decreased from $p = 0.012$ at the 1 km radius scale to $p = 0.037$ at the 4 km radius scale and 0.029 at the 5 km radius scale. The probability that an eagle owl nest would occur at spatial scales within a 1–5 km radius increased with the increase in the number of quarries and caves (Table 2). None of the mesohabitat cover types were significantly associated with the nest locations (Table 1). The most abundant land use surrounding the nests was planted fruit groves, which accounted for approximately 22% of the land cover. Planted woodlands were the next most common land cover, accounting for about 15–19%

of the land cover, and water bodies were the least abundant and accounted for <1% of the cover (Table 1).

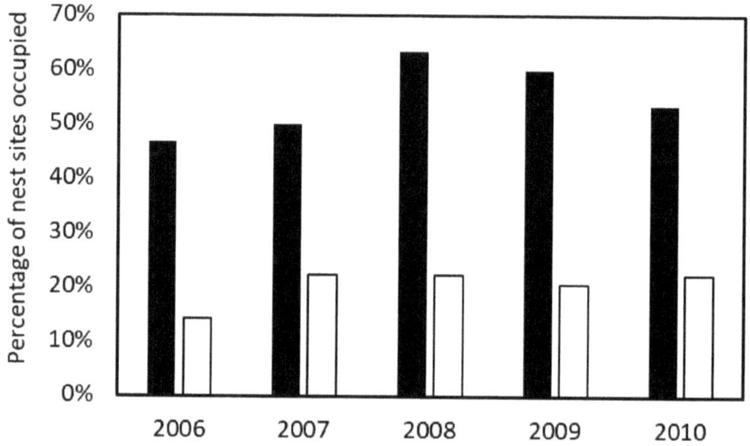

Figure 3. Comparison between the percentage of quarries (black, n = 30) and caves (white, n = 63) occupied by eagle owls (*Bubo bubo*) during 2006–2010. The preference for quarries is evident from the difference in nest placement between the two potential habitats.

Table 2. The probability of an eagle owl (*Bubo bubo*) nest in area (1–5 km radius) decreases with spatial scale, as expressed by lower significance values. Combined = number of caves and quarries. AIC denotes the Akaike Information Criterion.

Radius		AIC	Observed Mean	Simulated Mean	*p*-Value Model	*p*-Value Variable
1000	Cave	191.664	0.73	0.069	0.0001	0.0120
	Quarry	176.457	0.60	0.04	<0.0001	0.0004
	C + Q	153.781			<0.0001	<0.0001
2000	Cave	192.633	1.67	0.26	0.0002	0.0130
	Quarry	178.548	1.05	0.12	<0.0001	0.0005
	C + Q	159.405			<0.0001	<0.0001
3000	Cave	194.437	2.77	0.60	0.0005	0.0137
	Quarry	190.323	1.29	0.36	<0.0001	0.0006
	C + Q	175.112			<0.0001	<0.0001
4000	Cave	201.343	3.81	1.66	0.0234	0.0373
	Quarry	191.845	1.73	0.59	0.0001	0.0006
	C + Q	192.764			0.0002	0.0022
5000	Cave	200.991	5.45	2.42	0.0191	0.0290
	Quarry	190.916	2.18	0.85	<0.0001	0.0003
	C + Q	193.841			0.0004	0.0022

The probability that a specific nest site would be occupied during the study period (Table S1) and the number of nestlings fledged from the site were not significantly associated with the number of quarries and caves, nor the mesohabitat land around the nest site at any spatial scales (1–5 km radius; Table S2).

The linear mixed model ($F_{3,389}$ = 29.04, $p < 0.001$) including year, buffer group and buffer number as random factors found that the number of breeding pairs (mean = 0.73 active nests per buffer, range 0–7 nests) was positively correlated with the number of quarries ($F_{1,389}$ = 135.18, $p < 0.001$; mean = 0.50, range 0–6), number of caves (mean = 1.23, range 0–30; $F_{1,389}$ = 84.26, $p < 0.001$) and percentage of planted woodlands (mean = 517.75 Ha, range 0–3, 454.13 Ha; $F_{10,389}$ = 9.99, p = 0.002). They were not, however, related to the percentage of built-up areas ($F_{1,389}$ = 0.04, p = 0.85), disturbed areas ($F_{1,389}$ = 0.49, p = 0.48), grasslands ($F_{1,389}$ = 0.24, p = 0.63), maquis ($F_{1,389}$ = 0.05, p = 0.82), crop fields ($F_{1,389}$ = 11, p = 0.74), orchards ($F_{1,389}$ = 0.22, p = 0.64) nor water bodies ($F_{1,389}$ = 0.55, p = 0.50). Areas that had more quarries, caves and planted woodlands also had higher breeding densities.

4. Discussion

Our data suggested that the breeding population of the eagle owls in the Judea region of central Israel was limited by the lack of available nesting sites, i.e., quarries and caves. We found that the probability of an eagle pair breeding increased with the presence of both quarries and caves but was not related to mesohabitat land-use structure. Further, eagle owl breeding densities were related to the number of appropriate nest substrates and mesohabitat, mainly to planted woodlands. Furthermore, we found that eagle owls successfully raised their young regardless of the surrounding mesohabitat or the lack of additional nest sites around their nests, most likely due the owls' opportunistic and generalist hunting behavior which allows them to prey upon a varied prey base. Interestingly, our results concurred with [33], who concluded that cost–benefit evaluations regarding the cost of optimal-foraging, distance and height of nest sites on cliffs dictated site choice. Human disturbance was considered to be a prominent consideration in nest placement by eagle owls [9]. Our results concurred with some of the studies from Europe, where a shortage of nest sites was not found to affect breeding density [6,19].

The only mesohabitat structure that was related to the number of breeding pairs was the amount of planted woodlands, which was also related to the number of quarries, but not to the number of caves. This is most likely because woodlands were frequently planted around the former as part of environmental reclamation projects [34]. The major line of thought is that unused quarries should be rehabilitated because, when left untreated, they can cause land disturbances, as well as safety and environmental problems. The problem, as we see it, is that most of the rehabilitation is focused on how to convert the unused quarries into recreational areas for humans; the importance of biodiversity is not a priority [34]. Based on our findings, the number of potential nest localities is more important than the amount of planted woodlands, because the probability that an eagle owl nest was located at a site was significantly associated only with quarries and caves, but not with the amount of planted woodlands or recreational areas. Interestingly, planted woodlands may still be a significant factor in this issue since, in Spain, it was found that the amount of forest around the cliffs used by eagle owls as nest sites increased the probability of occupation [7], but decreased with human disturbance [9,12].

Unlike this study, where the occupation of nest sites was not related to the mesohabitat structures, the occupation of nest sites in France was positively related to open-habitat availability [6] and to protected areas in Spain [8]. In Europe, the number of fledglings was positively correlated to the percentage of open land [35], forested landscape [6] and protected areas [36], but was negatively correlated to the cover of urbanized areas [12] and proportion of wooded areas [14] around the nests. This underlines the fact that the mesohabitat around nest sites may affect breeding in ways that we have not yet elucidated. For example, the quality of nestlings (i.e., their body mass) could differ between habitats [37] and influence recruitment as well as population sustenance. Further studies are needed to determine whether the fitness and diet of owls may vary in the different mesohabitats.

Even though the eagle owl population in Israel is stable [38], the owls' dependence on quarries and caves for nest sites could be problematic locally, because the sites are not protected by law. As the human population of Israel is growing and habitat loss is becoming more prevalent [39], the carrying capacity and landscape continuity to sustain large predators, such as eagle owls, is impaired [40].

Furthermore, since eagle owls are large, attractive owls that breed in open areas on the ground and are easy to find, they are at risk of disturbance from recreational birders and wildlife photographers who frequently seek the owls out to watch and photograph them [41,42]. Anthropogenic disturbance was also found to affect nest-site occupancy by eagle owls in Spain; they bred less in nest sites located near paved roads [7,43]. Now that we have determined that the presence of nest sites encourage eagle owl breeding, it is vital to implement conservation practices to protect the quarries and caves from human disturbance or "development" exclusively for human purposes [33]. This is especially important in light of the fact that the bell caves are historical, human-created structures from ca. 2000 years ago and can only be maintained as historical sites [22]. Future studies are needed to assess the possibility of preserving quarries that are not necessarily within the boundaries of nature reserves, in order to sustain the present breeding population and to facilitate their dispersal into areas where the species currently does not breed.

The finding that a lack of appropriate nest localities may be a limiting resource to eagle owls in Israel may differ from other studies, due to the sampling effort. Here, we sampled a large area where eagle owls are known to breed. In addition, unlike Europe, Israel is a much smaller country with a denser human population, made up of habitat that is frequented by humans (i.e., from intensive agriculture, villages, recreational activities, etc.), and lacks large, extensive forest areas and wildlife refuges [40]. This study highlights the importance of protecting not only larger habitats, but also potential nest-site localities, from future land-use changes in the guise of development, even if only for recreational activities (i.e., rock climbing, mountain bicycles, off-road vehicles, etc.).

Even though eagle owls are generalists who breed in a wide variety of nesting habitats throughout many parts of their range [19], locations that limit nest sites highlight the complexity of wildlife conservation for apex predator species between regions and countries, especially in species with a wide global distribution such as the eagle owl. Inappropriate nesting sites for eagle owls appear to limit population numbers just like cavities can limit secondary cavity-breeders [5,44,45]. Hence, further studies are needed to determine if, by increasing artificial nesting sites, we could naturally influence dispersal and increase eagle owl numbers in areas with limited nest sites, but with a diverse and abundant prey base to sustain such populations [46]. This is especially important when taking into consideration that the study species is susceptible to human disturbance, while the human density in the region continues to grow. Multi-scale insights are required to ensure the persistence of the eagle owl populations in central Israel.

Supplementary Materials: The following are available online at https://www.mdpi.com/article/10.3390/d14060438/s1: Table S1. Analysis of land use—land cover on nesting success of eagle owls (Bubo bubo) at 1–5 km scales. "Gadash" denotes low crop fields. Table S2. Analysis of land use—land cover on mean annual number of fledglings, at 1–5 km scales. "Gadash" denotes low crop fields.

Author Contributions: E.H.: conceptualization and implementation; G.W.: GIS analyses, D.M. and M.C.: data analysis; D.M., M.C. and R.Y.: quality control and writing paper. All authors have read and agreed to the published version of the manuscript.

Funding: This research received no external funding.

Institutional Review Board Statement: The study did not require ethical approval and was conducted in the framework of the Israel Nature & Parks Authority.

Informed Consent Statement: Not applicable.

Data Availability Statement: Owing to the conservation sensitivity of the location of the nests, the data are available directly from the authors.

Acknowledgments: We thank Amran Tzabari, Yoram Shpirer, Tzipo Klein, Rea Shaish, and Irit Shahar for assistance in the field as well as Yasmin Raizner for assistance with the GIS. The paper publishing was paid using funding from Israel Ministry of Science and Technology to Motti Charter.

Conflicts of Interest: The authors declare no conflict of interest.

References

1. Brambilla, M.; Rubolini, D.; Guidali, F. Eagle Owl *Bubo bubo* proximity can lower productivity of cliff-nesting Peregrines *Falco peregrinus*. *Ornis Fenn.* **2006**, *83*, 20–26.
2. Fahrig, L. Non-optimal animal movement in human-altered landscapes. *Funct. Ecol.* **2007**, *21*, 1003–1015. [CrossRef]
3. McConkey, K.R.; O'Farrill, G. Cryptic function loss in animal populations. *Trends Ecol. Evol.* **2015**, *30*, 182–189. [CrossRef] [PubMed]
4. Newton, I. *Population Limitation in Birds*; Elsevier: Amsterdam, The Netherlands, 1998.
5. Newton, I. Population limitation in birds: The last 100 years. *Br. Birds* **2007**, *100*, 518–539.
6. Penteriani, V.; Gallardo, M.; Roche, P. Landscape structure and food supply affect eagle owl (*Bubo bubo*) density and breeding performance: A case of intra-population heterogeneity. *J. Zool.* **2002**, *257*, 365–372. [CrossRef]
7. Martínez, J.A.; Serrano, D.; Zuberogoitia, I. Predictive models of habitat preferences for the Eurasian eagle owl *Bubo bubo*: A multiscale approach. *Ecography* **2003**, *26*, 21–28. [CrossRef]
8. Ortego, J.; Díaz, M. Habitat preference models for nesting eagle owls *Bubo bubo*: How much can be interred from changes with spatial scale? *Ardeloa* **2004**, *51*, 385–394.
9. Donázar, J.A. Seleccion del habitat de nidificacion por el buho real (*Bubo bubo*) en Navarra. *Ardeola* **1988**, *35*, 233–245. (In Spanish, English Abstract)
10. Lourenço, R.; Santos, S.; Rabaça, J.E.; Penteriani, V. Superpredation patterns in four large European raptors. *Popul. Ecol.* **2011**, *53*, 175–185. [CrossRef]
11. Sergio, F.; Marchesi, L.; Pedrini, P. Integrating individual habitat choices and regional distribution of a biodiversity indicator and top predator. *J. Biogeogr.* **2004**, *31*, 619–628. [CrossRef]
12. Bionda, R.; Brambilla, M. Rainfall and landscape features affect productivity in an alpine population of Eagle Owl *Bubo bubo*. *J. Ornithol.* **2012**, *153*, 167–171. [CrossRef]
13. Donázar, J.A. Geographic variation in clutch and brood size of the Eagle Owl *Bubo bubo* in the Western Palearctic. *J. Ornithol.* **1990**, *131*, 439–443. [CrossRef]
14. Dalbeck, L.; Heg, D. Reproductive success of a reintroduced population of Eagle Owls *Bubo bubo* in relation to habitat characteristics in the Eifel, Germany. *Ardea* **2006**, *94*, 3–21.
15. Sergio, F.; Newton, I.P.; Marchesi, L. Top predators and biodiversity. *Nature* **2005**, *436*, 192. [CrossRef]
16. Sergio, F.; Marchesi, L.; Pedrini, P.; Penteriani, V. Coexistence of a generalist owl with its intraguild predator: Distance-sensitive or habitat-mediated avoidance? *Anim. Behav.* **2007**, *74*, 1607–1616. [CrossRef]
17. Tumurbat, J.; Sundev, G.; Yosef, R. Nest Site and Food Composition of the Eagle Owl *Bubo bubo* in Mongolia. *Ardea* **2009**, *97*, 519–523. [CrossRef]
18. Pande, S.; Pawashe, A.; Mahajan, M.; Mahabal, A.; Joglekar, C.; Yosef, R. Breeding Biology, Nesting Habitat, and Diet of the Rock Eagle-Owl (*Bubo bengalensis*). *J. Raptor Res.* **2011**, *45*, 211–219. [CrossRef]
19. Penteriani, V.; del Mar Delgado, M. *The Eagle Owl*; T & AD Poyser Ltd.: London, UK, 2019.
20. Clucas, B.; McHugh, K.; Caro, T. Flagship species on covers of US conservation and nature magazines. *Biodivers. Conserv.* **2008**, *17*, 1517–1528. [CrossRef]
21. Fischhendler, I.; Frumkin, A. Distribution, evolution, and morphology of caves in southwestern Samaria, Israel. *Isr. J. Earth Sci.* **2008**, *57*, 311–322. [CrossRef]
22. Zissu, B.; Kloner, A. The bell-shaped quarries of the Judean foothills, Israel. *Opera Ipogea* **2014**, *2*, 47–60.
23. Ratti, J.T.; Garton, E.O. Research and experimental design. In *Research and Management Techniques for Wildlife and Habitats*; Bookhout, T.A., Ed.; The Wildlife Society: Lawrence, Kansas, 1994; pp. 1–23.
24. Kozak, M. Optimal stratification using random search methods in agricultural surveys. *Stat. Transit.* **2004**, *6*, 797–806.
25. Andersen, D.E. Survey techniques. In *Raptor Research and Management Techniques*; Bird, D.M., Bildstein, K.L., Eds.; Raptor Research Foundation: Washington, DC, USA, 2007.
26. Penteriani, V.; Delgado, M.D.M.; Campioni, L. Quantifying space use of breeders and floaters of a long-lived species using individual movement data. *Sci. Nat.* **2015**, *102*, 21. [CrossRef] [PubMed]
27. Fleming, C.H.; Fagan, W.F.; Mueller, T.; Olson, K.A.; Leimgruber, P.; Calabrese, J.M. Rigorous home range estimation with movement data: A new autocorrelated kernel density estimator. *Ecology* **2015**, *96*, 1182–1188. [CrossRef] [PubMed]
28. Swindle, K.A.; Ripple, W.J.; Meslow, E.C.; Schäfer, D. Old-Forest Distribution around Spotted Owl Nests in the Central Cascade Mountains, Oregon. *J. Wildl. Manag.* **1999**, *63*, 1212–1221. [CrossRef]
29. Von Christoph, L. Telemetric study in the Eagl-Owl (*Bubo bubo*) in the foreland of the Alps in Lower Austria—Methods and first results. *Egretta* **1992**, *35*, 69–71.
30. Pande, S.A.; Yosef, R.; Deshpande, P.; Pawar, R.; Mahajan, M.N. Breeding biology of the mottled wood owl (*Strix ocellata*) in west-central inida. *J. Raptor Res.* **2018**, *52*, 240–244. [CrossRef]
31. Dalbeck, L.; Bergerhausen, W.; Krischer, O. Telemtriestudie zur Orts- und Partnertreue bei Uhu *Bubo bubo*. *Vogelwelt* **1998**, *119*, 337–344.
32. Raudenbush, S.W.; Bryk, A.S. *Hierarchical Linear Models: Applications and Data Analysis Methods*; Sage: Thousand Oaks, CA, USA, 2002; Volume 1.

33. Martínez, J.E.; Calvo, J.F. Seleccion de habitat de nidificacion por el buho real *Bubo bubo* en ambientes mediterraneos semiaridos. *Ardeola* **2000**, *47*, 215–220. (In Spanish, English Abstract)
34. Milgrom, T. Environmental aspects of rehabilitating abandoned quarries: Israel as a case study. *Landsc. Urban Plan.* **2008**, *87*, 172–179. [CrossRef]
35. Penteriani, V.; Gallardo, M.; Ferrer, M. Spatial heterogeneity and structure of bird populations: A case example with the eagle owl. *Popul. Ecol.* **2004**, *46*, 185–192. [CrossRef]
36. León-Ortega, M.; Martínez, J.E.; Pérez, E.; LaCalle, J.A.; Calvo, J.F. The contribution of non-protected areas to the conservation of Eurasian Eagle-owls in Mediterranean ecosystems. *Ecosphere* **2017**, *8*, e01952. [CrossRef]
37. Ortego, J. Consequences of Eagle Owl nest-site habitat preference for breeding performance and territory stability. *Ornis Fenn.* **2007**, *84*, 78–90.
38. Symes, A.; Taylor, J.; Mallon, D.; Porter, R.; Simms, C.; Budd, K. *The Conservation Status and Distribution of the Breeding Birds of the Arabian Peninsula*; IUCN: Cambridge, UK; Gland, Switzerland; Environment and Protected Areas Authority: Sharjah, United Arab Emirates, 2015. [CrossRef]
39. Orenstein, D.E.; Hamburg, S.P. Population and pavement: Population growth and land development in Israel. *Popul. Environ.* **2010**, *31*, 223–254. [CrossRef]
40. Levin, N.; Lahav, H.; Ramon, U.; Heller, A.; Nizry, G.; Tsoar, A.; Sagi, Y. Landscape continuity analysis: A new approach to conservation planning in Israel. *Landsc. Urban Plan.* **2007**, *79*, 53–64. [CrossRef]
41. Tipling, D. Bird photography—A new code of practice. *Br. Birds* **2011**, *104*, 223–224.
42. Radulski, B.G. Understanding Human Disturbance to Birds at the Intersection of Birding and Bird Photography. Ph.D. Thesis, Purdue University Graduate School, West Lafayette, IN, USA, 2020.
43. Martínez, J.E.; Martínez, J.A.; Zuberogoitia, I.; Zabala, J.; Redpath, S.M.; Calvo, J.F. The effect of intra- and interspecific interactions on the large-scale distribution of cliff-nesting raptors. *Ornis Fenn.* **2008**, *85*, 13–21.
44. Iezekiel, S.; Yosef, R.; Bakaloudis, D.E.; Papakosta, M.A.; Vlachos, C.G.; Antoniou, A.; Zduniak, P. The endemic Cyprus Wheatear (*Oenanthe cypriaca*) adapts readily to artificial nest sites. *Biol. Conserv.* **2017**, *213*, 1–4. [CrossRef]
45. Iezekiel, S.; Yosef, R.; Themistokleus, C.; Bakaloudis, D.; Vlachos, C.; Antoniou, A.; Iezekiel, E.; Papakosta, M.; Kosicki, J. Endemic Cyprus Scops Owl *Otus cyprius* Readily Breeds in Artificial Nest Boxes. *Animals* **2021**, *11*, 1775. [CrossRef]
46. Crooks, K.R.; Soulé, M.E. Mesopredator release and avifaunal extinctions in a fragmented system. *Nature* **1999**, *400*, 563–566. [CrossRef]

MDPI
St. Alban-Anlage 66
4052 Basel
Switzerland
www.mdpi.com

Diversity Editorial Office
E-mail: diversity@mdpi.com
www.mdpi.com/journal/diversity

www.ingramcontent.com/pod-product-compliance
Lightning Source LLC
Chambersburg PA
CBHW061618210926
51850CB00009B/2361
9783036586922